THE FIRST FLEETS

MARITIME CURRENTS: HISTORY AND ARCHAEOLOGY

SERIES EDITOR

Gene Allen Smith

EDITORIAL ADVISORY BOARD

John F. Beeler
Alicia Caporaso
Annalies Corbin
Ben Ford
Ingo K. Heidbrink
Susan B. M. Langley
Nancy Shoemaker
Joshua M. Smith
William H. Thiesen

The FIRST FLEETS

COLONIAL NAVIES OF THE BRITISH ATLANTIC WORLD, 1630–1775

BENJAMIN C. SCHAFFER

The University of Alabama Press
Tuscaloosa

The University of Alabama Press
Tuscaloosa, Alabama 35487-0380
uapress.ua.edu

Copyright © 2025 by the University of Alabama Press
All rights reserved.

Inquiries about reproducing material from this work should be addressed to the University of Alabama Press.

Typeface: Alegreya

Cover image: View of the English landing on the island of Cape Breton to attack the fortress of Louisbourg, 1745; engraving by F. Stephen
Cover design: Lori Lynch

Cataloging-in-Publication data is available from the Library of Congress.
ISBN: 978-0-8173-2234-2 (cloth)
ISBN: 978-0-8173-6206-5 (paper)
E-ISBN: 978-0-8173-9559-9

I think a circumstantial history of naval operations in this Country ought to be written even as far back as the [1750s] province ship under Captain Hollowell, &c., and perhaps earlier still.

—Former president John Adams, in a letter to former vice president Elbridge Gerry, describing the origins of the US Navy

Contents

Acknowledgments ix
Glossary of Naval Terminology xiii
A Note on Early Modern Dates and Historical Grammar xvii

Introduction 1
1. The Transatlantic Roots of Provincial Navies up to 1685 8
2. Provincial Navies in the First Imperial Wars of 1689–1713 27
3. Provincial Navies and Irregular Warfare on Imperial
   Borderlands, c. 1713–1739 48
4. The War of Jenkins' Ear and the Incomplete "Royalization"
   of American Naval Defense 82
5. The Replacement of Provincial Navies by the Royal Navy in
   the Seven Years' War, c. 1756–1763 111
6. The Legacy of Provincial Navies and the Navies of the American
   Revolution, c. 1762–1775 128
Conclusion 144

Notes 147
Bibliography 183
Index 207

Illustrations follow page 75.

Acknowledgments

I am indebted to so many individuals for this book's publication. Firstly, I would like to thank Dan Waterman and the publication/editorial staff at the University of Alabama Press for their support throughout the publication process. I would also like to thank Dr. Gene Smith for first connecting me with the university press. I'm also thankful for the vast community of academic historians, scholars, and archivists at the University of New Hampshire, the College of Charleston, Notre Dame University, and the Charleston County Public Library. Chief among this list, I would like to thank Dr. Eliga Gould for mentoring me throughout my graduate and postgraduate studies, for introducing me to the study of Atlantic world history, and for supporting my scholarship both in person and virtually. I would also like to express gratitude to Drs. Cynthia Van Zandt, Jessica Lepler, David Bachrach, Patrick Griffin, Mike Verney, Phyllis Jestice, Christophe Boucher, Lisa Covert, Jacob Steere-Williams, Sandra Slater, Jason Coy, and Nic Butler for their generous advice, wisdom, encouragement, editorial assistance, moral support, and feedback throughout the research, writing, and publication process. Furthermore, I would like to thank the graduate school at the University of New Hampshire, the Nguyen family, and the Steelman family for their generous financial support and fellowships that allowed me to travel to the UK and Washington, DC, to examine vital primary sources in person.

In my personal life, I deeply appreciate the support and enthusiasm from my friends in the living history community, and Vicar Callie Walpole of St. John's Episcopal Church. Furthermore, I would also like to acknowledge the loving support of my family, including those who supported my early interest in history but sadly passed before they could see this book published (my late grandparents Bunky and Joyce Williams, aunt Pam Hansford, and adopted uncle Michael Kogan). Finally, and perhaps most importantly, I thank my parents, Fred and Sandy Schaffer. My parents have not only lovingly supported my research efforts but have always encouraged me and reminded me that—with a firm reliance on Divine Providence—I should always follow my love for history wherever it takes me.

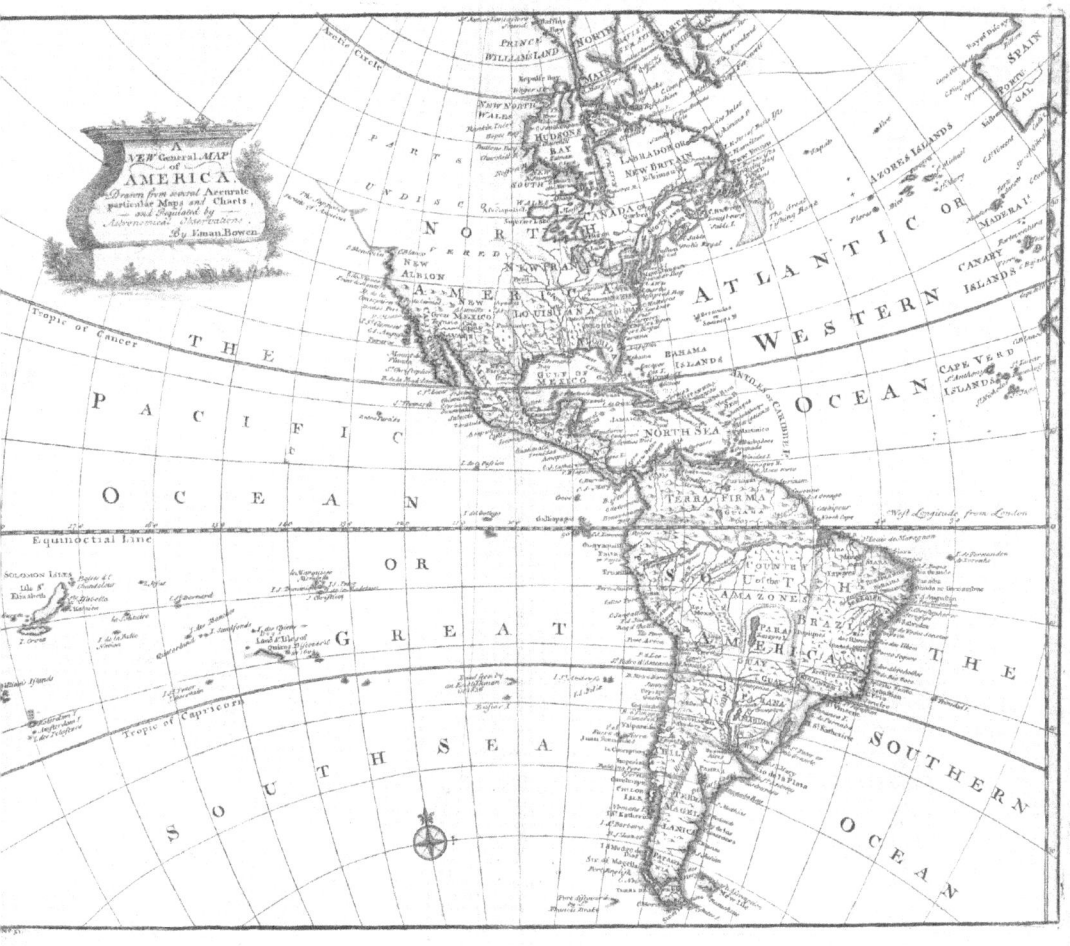

FIGURE 1. E. Bowen, *A complete system of geography. Being a description of all the countries, islands, cities, chief towns, harbours, lakes, and rivers, mountains, mines, &c. of the known world* [. . .], 1747 edition. While most of the regions consulted in this study are visible in this map, Halifax, Nova Scotia, is not depicted as it would not be established until 1749. Wikimedia Commons.

FIGURE 2. *A view of Charles Town the Capital of South Carolina.* This eighteenth-century depiction of Charles Town's waterfront includes many of the vessels that were common in provincial navies. On the far left is a typical single-masted trading sloop. Just behind the sailors in the foreground is a very small coastal sailing vessel, similar in structure to the periagua sailing canoes of the South Carolina provincial navy. In the center is a fully rigged ship, similar to British Royal Navy frigates. The two-masted vessel to the right of the ship is likely a coastal trading schooner. New York Public Library Digital Collections.

Glossary of Naval Terminology

The following glossary will give readers a brief guide to the types of vessels referred to in this book. Most provincial Anglo-American maritime craft were built for mercantile trade and were not purpose-built warships like their compatriots in the British Royal Navy. These vessels were often classified by their "rig" or the ways in which the sails were arranged. For instance, a "fore-and-aft" rigged vessel has a sail pointing toward the front (fore) of the vessel to the back (aft). The term "ship" usually referred to a "ship-rigged" vessel with three large masts with square sails. From the fore to the aft of a ship, the masts of a ship were known as the "foremast," "mainmast," and "mizzen mast." Sails were suspended from "yards" on these masts.

Brigantine: Brigantines were typically small vessels under one hundred tons. They had two masts. The foremast was square rigged while the mainmast had a fore-and-aft rig. This differed slightly from a brig, which was a two masted vessel that typically had square-sail rigs on both of its masts.[1] After sloops, brigantines may have been the most common merchant vessel employed in provincial fleets.

Frigate: A large, ship-rigged warship usually belonging to the Royal Navy and, in very rare cases, provincial navies. Typically, Royal Navy frigates sent to America were among the smallest "rated" warships. While first- or second-rate warships with nearly one hundred guns would be reserved for European service, smaller fifth- or sixth-rate frigates that carried between twenty and forty guns would serve in the New World.[2]

Galley: The term "galley" has been used to describe light-drafted rowing vessels since classical antiquity. In the British Atlantic world, the term had various meanings. Occasionally, a vessel named *X Galley* would simply be a regular trading vessel that had a hull specifically shaped for swifter sailing. On the other hand, some colonial governments particularly in the Southeast built light-draft "galleys" that relied on oars as their primary source of propulsion.[3]

Ketch: A very small ocean-going trading or fishing vessel with a main mast and mizzen mast but no foremast. In New England, these vessels were typically under seventy tons.[4]

Periagua/Periauger/Piragua: Inspired by Native American dug-out canoes, periaguas (spelled and pronounced in myriad ways) were small, swift vessels similar to galleys in that they were primarily powered by oars and occasionally by sails. These vessels were particularly common in the southern colonies and the West Indies and were the main vessels of the South Carolina Scout Boat navy.[5]

Pink: A vessel with a "pinked stern" or rear of a ship that was narrow when compared to the rest of the vessel.[6]

Pinnace: This term had multiple meanings in the seventeenth century. In essence, this seems to have been a catch-all term for many varieties of smaller vessels used for coastal operations, although larger pinnaces of one hundred or more tons burthen were not unheard of. One 1686 source described a pinnace as a small craft with a square-shaped stern that could be powered by sails on three masts and oars.[7]

Schooner: Vessel with two masts that were rigged in a fore-and-aft pattern. Square sails could be added on top of the masts to make them topsail schooners. They were somewhat similar in appearance to brigantines but had narrower hulls, and their masts were more slanted.[8]

Shallop: Typically, a very small, open-decked coastal work boat. Larger vessels often times carried shallops onboard to serve as auxiliaries when needed.[9]

Sloop: The most common vessel in provincial navies and perhaps the most commonly employed vessel in the British Atlantic world. In the North American (and provincial navy) context, a sloop usually described a single-masted trading vessel with a fore-and-aft rig that could involve any number of sail types. These smaller vessels were typically under one hundred tons and could operate in shallow waters off the coasts or on ocean-going missions.[10]

Snow: This vessel was extremely similar to a brig except for the fact that its rear mast (known as the "trysail-mast" due to its prominent fore-and-aft trysail that jutted out of the rear of the vessel) was close to its mainmast. The 1740s Massachusetts province snow *Prince of Orange* was a prime example of this vessel type.[11]

GLOSSARY OF NAVAL TERMINOLOGY   xv

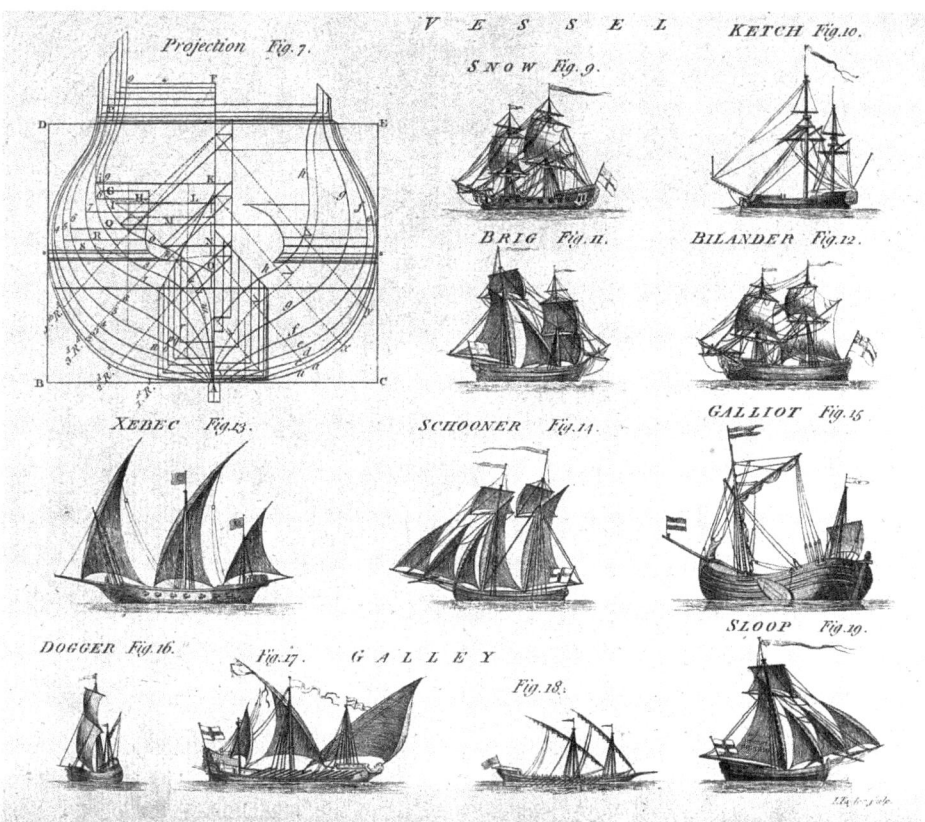

FIGURE 3. Eighteenth-century diagram of various vessel types common throughout the Atlantic world: brig, billander, dogger, galley, galliot, schooner, sloop, xebec. Engraving by J. Taylor. Source: Wellcome Collection Online.

Whaleboat: Whaleboats were extremely common small craft employed primarily by New England mariners for whale hunting but were also used throughout the colonies for various military missions. They were particularly useful for troop transport and could be powered by oars or sails.[12]

A Note on Early Modern Dates and Historical Grammar

My goal in this book has been to preserve the integrity of seventeenth- and eighteenth-century sources while also remaining accessible to the modern reader. With this in mind, I have automatically modernized the early modern English spellings of the words *ye*, *yt*, and *ym* to *the*, *that*, and *them*. I have also substituted whole words for words that are put in superscripts. In the seventeenth and eighteenth centuries, there was a tendency to irregularly shorten words like *general* to *gen^ll*. I have modernized these spellings in all cases—even if primary sources printed in modern books or articles maintained the now obsolete spellings. I have also modernized words that make use of the now-obsolete long *s* (ſ). Aside from modernizing letters, I have not replaced seventeenth- or eighteenth-century manners of spelling or inconsistent capitalizations. Despite the best efforts of autocorrect, I believe it would be anachronistic to completely use twenty-first century spelling conventions for early modern English. In other words, a sentence that was originally written "Y^e Sailors tooke a Boate to go Fiſhing" will read as "The Sailors tooke a Boate to go Fishing." The original feel of the primary source remains with some alterations to ease the experience of modern readers.

Despite these modernizations, I have also not altered the Julian (a.k.a. Old Style) dating system. Before 1752, the British Empire—unlike much of the rest of Europe—still used the antiquated Julian Calendar, which ran between ten and eleven days behind the more popular Gregorian Calendar. That calendar marked the New Year on March 25 (Lady Day), and from January 1 to March 24 used a "double date" that marked the transition between years (e.g., January 15, 1709/10). To eradicate this dating system would prove confusing for individuals who wish to investigate the sources cited in this book, and this system of dating has been maintained for this reason.

THE FIRST FLEETS

Introduction

Philadelphia was in danger. It was the autumn of 1747—the eighth year of the third global imperial war Britain had fought against its Franco-Spanish enemies in half a century—and French and Spanish privateers prowled nearby in Delaware Bay. While the hundreds of thousands of Anglo-American colonists between Newfoundland and Barbados faced occasional terrestrial threats from hostile French, Spanish, and Native American forces, the largely coastal British American colonies and their ocean-bound commerce suffered even more so from this Franco-Spanish commerce raiding. In the face of this imminent threat, the Pennsylvania legislature struggled to find a solution to protect vulnerable local merchant ships.

While debates over coastal defense measures were common in every British province, the Quaker-dominated proprietary colony of Pennsylvania was unique. On the one hand proprietary colonies were essentially American fiefdoms that were privately owned by absentee landlords living in England and thus were never guaranteed royal military protection that colonies directly under the auspices of the Crown (e.g., Virginia) enjoyed.[1] On the other hand, Pennsylvania's governors regularly struggled to convince the pacifistic and tight-fisted Quakers that dominated the Pennsylvania Assembly to expand the colony's naval defenses.[2]

What military measures could a colony without Royal Navy protection take to combat enemy privateers? The colony's government could fund and direct the construction of a local *provincial navy*. As early as 1634, the infant Massachusetts government pioneered this tactic by fitting out a guard ship to hunt for pirates with local resources and men.[3] With Royal Navy involvement in North America and the West Indies limited before the second decade of the eighteenth century, Anglo-American governments would frequently use local funds and sailors to build regional defense fleets to protect ports and commerce from enemy navies and pirates.[4] On occasion, these provincial fleets could also be deployed in offensive campaigns against enemy port cities. While the Anglo-American precedent for provincial navies extended back to the beginning of the seventeenth century, colonial governments found even more reason to build these local fleets when naval threats amplified during four global conflicts between Britain and its imperial/Native foes: King William's War (1689–98), Queen Anne's War (1702–13), the War of Jenkins' Ear/King George's War (1739–48), and the Seven Years' War (1754–63). With each ensuing conflict, the North American and West Indian provinces were drawn more and more into deadly battles for maritime hegemony in the New World.

So, in Philadelphia in 1747 we find the forty-one-year-old printer and politician Benjamin Franklin—a future founder of both the United States and its first fleet, the Continental Navy—arguing for Philadelphians to support a local warship to hunt down Franco-Spanish privateers. In a pamphlet entitled *Plain Truth*, Franklin warned that the "Absence of [Royal Navy] Ships of War, during the greatest Part of the Year, from both Virginia and New-York" left the port city vulnerable to maritime assaults. To secure their coast, Philadelphians should work together to bear the "Expence of a Vessel to guard our Trade."[5]

Similar debates occurred throughout much of the British Atlantic world when enemy privateers lurked off the coast, when pirates captured local merchant ships, or when the Crown called on colonial governments to initiate assaults on enemy port cities. Constructing and financing a local navy was no small task for pre-Revolutionary provincial governments, and the manner in which provincial navies were created and funded varied from colony to colony. When local officials resorted to unpopular measures such as impressment of vessels and sailors or when they instituted burdensome taxes to fund local defense measures, they risked igniting the potentially violent anger of the populace. Even when colonial governments successfully established a temporary or semipermanent defense fleet, the associated costs often times aggravated already-potent internal sociopolitical tensions throughout the American colonies.

If the organization of provincial navies elevated tensions within Anglo-American colonies themselves, they also raised larger questions over the provincial-royal relationship. Was the Crown or its colonies responsible for coastal security? Who would pay for provincial ships? Were American provincial captains equal partners to Royal Navy captains or subordinates?

Who would man Royal Navy frigates in American waters? What could colonial governments do if royal captains committed crimes or did not actively patrol for enemy vessels? These questions were never adequately answered in the pre-Revolutionary era and numbered among the myriad cracks in the relationship between periphery and center that would shatter in the imperial crisis of the 1760s and 1770s.

In this book, I make the case that between the 1630s and 1740s, a weak and disorganized Royal Navy presence in the New World led Anglo-American governments to build and utilize their own provincial navies against French, Spanish, Native, and piratical maritime threats. By the mid-eighteenth century, the British government finally expanded the Royal Navy's presence throughout the Atlantic world. Nevertheless, provincial anger at perceived excesses by Royal Navy captains hampered these piecemeal attempts to "royalize" coastal defense and contributed to the growing rift between Britain and its colonies. During the imperial crisis of the 1760s–70s and ensuing Revolutionary War, this anger—coupled with the long-ingrained tradition of provincial naval defense—would play a major role in shaping America's resistance to British authority at sea.

The "Privateering" Conundrum and Historiography

In its broadest historiographical sense, *The First Fleets* attempts to correct a glaring omission in early American and maritime scholarship: scholars have generally ignored the fact that Anglo-American governments maintained their own navies throughout the century leading up to the American Revolution. I do not believe this omission has come from a lack of interest or intentional suppression of sources but from the mistaken assumption that there were only two sorts of maritime forces active along the coasts of colonial America: the British Royal Navy and *privateers* (commonly defined as crews of private warships hired or licensed by local governments). The prevalence of this belief is evident when scholars make statements such as "the colonies did not have permanent armies or navies, and there was not even a maritime equivalent to the rudimentary military training provided by colonial militias."[6] While this book contests this assertion and aims to bring attention back to the oft-ignored story of provincial naval warfare, we must first discuss its relationship to the historically and historiographically hazy concept of privateering.

This "haziness" can be traced back to the vast array of different types of naval organizations present in the medieval and early modern worlds.

Historian N. A. M. Rodger argues that when one casts aside modern notions of navies being "permanent fleets of warships, manned by professional officers and men, supported by an elaborate infrastructure and maintained from the revenues of central government," they can find records of a diverse array of naval forces in medieval and early modern Europe. These forces included fleets of vessels impressed or commissioned by monarchs, feudal or regional fleets raised by local lords and magnates, and various sorts of "private" naval forces (nongovernmental commercial vessels that waged wars against rivals, with or without government approval). Rodger argues that when the Spanish Crown forbade any other European powers from accessing the riches or trade of the New World in the sixteenth century, England and other northern European kingdoms encouraged private commercial warfare against the Spanish throughout the Atlantic world. This meant that irregular private naval warfare was "artificially preserved [in the Americas] long after it had disappeared from European waters."[7]

It was from "private naval warfare" that Rodger asserts that the seventeenth-century term "privateer" originated. Up to the seventeenth century, private merchant ship crews often armed themselves and fought defensive actions against enemy raiders, took part in occasional pirate raids, or sought out permission for "reprisals" from their monarchs to retrieve stolen property. By the sixteenth century, various European monarchs—plagued by constant religious warfare and their own lack of warships—began to license private warships to raid enemy commerce for profit. Despite these early cases of government-sanctioned commerce raiding, it took the English Crown until the late seventeenth century to fully codify the state's role in private naval warfare and, more specifically, to use the term "privateer" to describe privately commissioned commerce raiders.[8]

In my own research, I have found Rodger's lament that "generations of scholars have made difficulties for themselves and their readers by using vague, anachronistic and contradictory language about private naval warfare" to ring true.[9] In particular, I have found that the definition of the word "privateer" has only grown more expansive and vaguer throughout the centuries. One 1720 dictionary defined a privateer as "a Vessel fitted out by one or more private Persons, with a Licence from the Prince or State, to prey upon the Enemy; also the Commander or Captain of such a Ship." A few decades later, British writer Samuel Johnson defined a privateer as a "a ship fitted out by private men to plunder enemies. He is at no charge for a fleet, further than providing *privateers*, wherewith his subjects carry on a piratical war at their own expence."[10] In both of these contemporary definitions, privateering was seen as an independently controlled activity with tacit government acceptance.

For the most part, provincial governments in the seventeenth and eighteenth centuries followed the aforementioned definitions, using the term "privateer" or "private men of war" to describe privately licensed commerce raiders. When describing war vessels fitted out by colonial governments, they typically used terms like "sloop of war," "province sloop," "vessel fitted out at the expense of the government," and so on. This was not always the case, and I have certainly found some cases where provincial authorities built warships and called them privateers. Nevertheless, the prominent Anglo-American insistence that tax-funded provincial fleets were something more than privateers seems to have grown over the decades as these fleets grew in complexity. As will be seen in chapter 3, the battle over what constituted a "privateer" or a "warship" led to a transatlantic legal battle between provincial and Royal Navy captains in the 1740s.

If privateering was an ill-defined term in the colonial era, later historians have done nothing to narrow its categorical grasp. As early as the mid-1920s, historian Howard Chapin argued that privateer ships were "privately owned armed-vessels, which sailed under the flag and commission of some recognized government." Chapin also maintains that by the 1700s, privateers included both sailors—who mainly chose to attack enemy shipping with legal permission ("privateers")—and merchants—who occasionally exercised the right as a sort of side job ("letter-of-marque ships"). Chapin includes in this "privateering" category colonial government-owned ships and even ships impressed for emergency reasons.[11]

Despite Chapin's inclusion of these government-directed naval activities within the scope of "privateering," one begins to wonder how "private" a government-directed expedition could be. The vagueness of Chapin's handling of "privateering" becomes especially apparent when he calls New England governor Edmund Andros's privateering fleet the "beginnings of a colonial navy."[12] Historian Charles O. Paullin complicated the definition of "colonial navies" in the next decade. Where Chapin groups government-sponsored naval expeditions in with privateering, Paullin separates the two with great nuance. In his unpublished 1930s manuscript, *Colonial Army and Navy*, Paullin argues that "the war vessels of the American colonies were of two general classes: (1) vessels under the control of the state, and (2) privateers. The former were of three classes: (1) vessels owned by the state, (2) vessels hired by the state, and (3) vessels freely loaned to the state." For Paullin, privateers were sailors given private commissions to pursue enemy commerce, while colonies maintained their own "war vessels." Unlike Chapin, Paullin extends his examination of Anglo-American fleets well into the Seven Years' War and notes increasing complexity and naval organization in some cases throughout various colonies.

Despite his nuanced handling of the different Anglo-American naval forces throughout the pre-Revolutionary era, Paullin's unpublished account seems to be a mere rough draft and concludes on a very questionable claim. While admitting that colonial fleets could barely be called proper "navies" at all and that Massachusetts had something of a "rudimentary navy" in the mid-eighteenth century, Paullin concludes that Anglo-Americans were "practically without a naval defense, except such as could be extemporized in emergencies." He also contends that from 1690 "they had the protection afforded by a few ships of the Royal Navy; and potentially of course, they, being a part of the British empire, were defended by the whole British Navy."[13] As I discuss in this book, the Royal Navy was anything but dependable in the eyes of Anglo-American governments.

Later in the twentieth century, colonial military and naval historians such as W. A. B. Douglas, Larry Ivers, and Carl Swanson used terms such as "sea militias," "coast guards," and "provincial navies" to differentiate between provincial government-funded warships and commerce raiders with letters of marque.[14] While these scholars mentioned provincial naval forces in their studies—and even made a point of separating them from forces of privateers—the importance of state-run maritime forces throughout the Atlantic world has still not yet received a study of its own.

All in all, with few exceptions, historians over the last century have largely only mentioned provincial navies in passing or have grouped them into the ever-growing and murky category of privateering. While admitting the blurred lines between colonial navies and privateers, with this book I propose to resurrect elements of Chapin's and Paullin's early twentieth-century categorizations of provincial navies. In this study, I define a *provincial navy (or provincial navy vessel)* as a war vessel or group of war vessels directly funded and outfitted by colonial American governments. These vessels could be guard ships used to protect commerce, temporarily impressed flotillas in emergencies, or hastily assembled invasion fleets that colonial authorities armed to attack enemy ports. In most cases, these vessels were primarily small merchant vessels that were taken into provincial service rather than purpose-built warships.[15]

One could be forgiven for thinking that much of this discussion over terminology is handwringing over semantics. After all, what difference does it really make if a historian calls a fleet of colonial war vessels operated by provincial governments "privateers"? By categorizing provincial naval establishments as merely collectives of privateers, historians ignore the fact that Anglo-American governments possessed the economic and political capital to maintain offensive and defensive fleets *generations* before the American Revolution—the moment often pointed to as the birth of the American naval tradition. In essence, the frequency with which Anglo-American governments built and maintained complex naval establishments speaks to a larger colonial capacity for state building and sea power in the decades leading up to the American Revolution than has previously been appreciated.[16]

While *The First Fleets* challenges prevailing scholarly conceptions of colonial naval defense, it also engages with broader scholarly discussions on war and society in early America. On the one hand, this book brings a new perspective to long-standing debates over the extent of the British imperial center's control over its American peripheries and highlights the agency of Anglo-American military forces and governments in shaping Britain's military strategies in the New World.[17] On the other hand, *The First Fleets* also joins a growing wave of scholarship that emphasizes the role of maritime conflict in shaping and challenging the diverse peoples, cultures, institutions, and societies of the pre-Revolutionary Atlantic world.[18]

On a final note, while this book investigates imperial political dynamics and governmental military planning, it is by no means an old-school military history that only concerns itself with topics of grand strategy and politics. Throughout *The First Fleets*, I constantly urge readers to consider the role of common sailors, maritime communities, and marginalized peoples in the wider story of provincial naval warfare. *The First Fleets* is not only the story of the earliest Anglo-American fleets that defended their shores in the decades prior to the Revolution but also of the maritime communities and sailors that struggled, fought, and bled to keep them afloat.

1

The Transatlantic Roots of Provincial Navies up to 1685

In the frigid late autumn of 1632, John Winthrop—governor of the newly established Massachusetts Bay colony—received a warning from officials in New Hampshire that a renegade English pirate named Dixie Bull, along with a crew of fifteen men, was pillaging vessels along the New England coast. Those officials had sent out four small pinnaces and shallops—manned by forty sailors—to find the pirate and hoped that their compatriots to the south would help with the hunt. Winthrop summoned a council and agreed to send his own bark "with twenty men to join with those [of Piscataqua, New Hampshire] for the taking of the said pirates." Wintry weather impeded the search and ultimately Bull escaped from capture.[1]

Three centuries later, an early twentieth-century writer posited that the joint hunt for Bull was the "first hostile fleet fitted out in New England and the first naval demonstration made in the colonies."[2] While the fitting out of an ad hoc emergency fleet of local vessels was still a novel activity in the infant English colonies of the New World, Winthrop and his compatriots were actually continuing a largely medieval tradition of localized naval defense. At the dawn of English colonization in the New World in the early seventeenth century, the notion of a centrally controlled Royal Navy with hundreds of state-owned warships that "ruled the waves" was a facet of a distant future.

Throughout the seventeenth century, local naval defense throughout the English colonies of the New World relied on familiar private and regional maritime defense strategies from the Old World. In an era when the English government still largely depended upon the naval power of private shipowners, the first generation of English colonists in the New World continued the ancient English trend of locally organized naval defense. Even as the Royal Navy began a slow path toward professionalization throughout the seventeenth century, regional provincial navies along with privateering became the primary methods of naval defense in the fledgling English Atlantic world.

State and Private Naval Forces of Medieval England, c. 1000–1400

The story of Anglo-American provincial navies necessarily begins in the homeland of most of those earliest colonists—England. Broadly speaking, English monarchs in the Middle Ages had neither the interest nor the capability to command a centralized or permanent navy of any import. In fact, in recent years, maritime historians have largely concluded that strict legal definitions of early modern concepts such as "standing navies," "privateering," or "piracy" would have been foreign to medieval Englishmen and continental Europeans alike.[3] Nevertheless, historian N. A. M. Rodger's useful categorization of medieval and early modern naval warfare as belonging to "two broad categories . . . public and private, or military and commercial" will be utilized here.[4]

While eighteenth-century English monarchs would one day utilize the Royal Navy to defend commerce from maritime predation, their medieval predecessors did not consider the defense of merchant ships in the pirate-infested waters of the English Channel to be a royal responsibility. In a time period when "few if any medieval ship-owners could expect to make money without being willing to fight for it," it is unsurprising that the onus of the defense of commerce fell to merchant captains themselves. If the defense of commerce was a low priority for royal authorities in the Middle Ages, the question necessarily arises: When and how did the English Crown organize what we might call navies? By and large, monarchs from Edgar the Peaceful in the tenth century to James I in the seventeenth century typically relied on flotillas of vessels provided by their subjects to transport soldiers and equipment to battlefields on the European continent, to repel invasions, and to engage enemy navies during times of war.[5]

As far back as Viking predations against Anglo-Saxon England, English monarchs depended on a small number of their own privately owned vessels and larger collectives of their subjects' requisitioned or hired vessels for naval expeditions. During the late tenth century, King Edgar of Wessex began to institute a naval taxation system known as the "ship-soke" tax. The Crown required large landholders to provide a certain number of ships and soldiers to the state based on the size of their domains. By the early eleventh century, the ship-soke system likely provided several hundred vessels for King Aethelred II's campaign against Danish Vikings.

By the mid-eleventh century, King Edward the Confessor shaped his naval forces in light of the well-established *fyrd* system. The *fyrd* was a kingdom-wide levy of able-bodied foot soldiers—a system that inspired what seventeenth-century Anglo-Americans would one day call the militia. When needed, Edward's subjects not only provided vessels via a ship-tax system but also fitted these vessels out with soldiers and sailors known as the *scipfyrd*. Regional fleets utilized by the Crown mustered in London and were typically supplied and manned by their respective home regions for temporary service. If naval forces were needed for longer periods, Crown officials rotated crew members every few months to prevent long periods of overemployment. Aside from regional musters of vessels, during the eleventh century the Crown also occasionally hired foreign mercenary vessels to patrol the English coasts and also began to offer special legal privileges to various towns in Kent that provided a permanent fleet of war vessels to scout for enemies in the English Channel. These ports, including Cornwall and Dover, would later form a trading collective known as the Cinque Ports.[6]

Royal reliance on private naval resources did not vary for much of the following three centuries. To illustrate this point, the Crown only owned fourteen ships out of a fleet of 331 war ships assembled by the English government in 1338. Most of the other vessels were likely impressed by royal officials and temporarily seized from their owners. In this period, ship impressment began after the king issued a general embargo on all merchant vessel departures. Royal officials, typically from the office of the exchequer, along with various maritime experts surveyed vessels and handpicked vessels appropriate for either carrying cargo and men or for combat. The Crown often depended on shipowners themselves to feed and pay for the impressment process—an expectation that proved to be economically challenging for many shipowners. When expeditions began, the Crown often paid for the personnel on these ships, but financial compensation for shipowners regularly came long after the expeditions and often in the form of partial arrears.

Royal impressment policies and demands for regional supplies of ships could be successful during times of broad public support for conflicts. Nevertheless, counties, cities, and individual shipowners often opposed royal mandates when the hardships of war became too burdensome. Sailors sometimes refused to sail without advanced payments, ship captains escaped from impressment embargos, and angry mobs even attacked royal officials and purposefully delayed ships. Due to economic privations created by this system and popular resistance, by the late fourteenth century, the Crown and Parliament began to slowly offer more regular reimbursement to shipowners and crews.[7] This sort of grassroots resistance by sailors to maritime impressment policies would frequently reemerge throughout the Atlantic world in the seventeenth and eighteenth centuries.

There were moments during the Middle Ages when the Crown attempted to keep more regular standing naval forces at its disposal. For instance, in the first quarter of the fifteenth century, during the Hundred Year's War against France, King Henry V purchased various ships to patrol the English Channel for enemy shipping. Rather than setting a precedent for a future centrally controlled Royal Navy, however, Henry V's fleet was a short-lived experiment. Due to rising debts from the long conflict with France, his heir, Henry VI, sold the fleet in the 1420s to alleviate the growing cost of a standing navy.[8]

Ultimately, the failure of Henry V's "standing" navy speaks to the immature state of naval administration in late medieval England. Royal ships were overseen by a single official known as the Clerk of the King's Ships. The clerk—who may or may not have had any actual naval experience—had a variety of poorly defined duties ranging from managing naval stores and ship repairs to paying the masters of the vessels. Beyond this broad managerial position, there was not yet any permanent Admiralty or navy boards as the state of the small royal fleet did not yet require such bureaucratic organizations.[9]

While we have considered Crown-controlled naval forces thus far, it is also important to note that medieval naval warfare depended as much—if not more—on private initiative. During the Middle Ages, maritime trade was a violent, competitive, and risky business. This was a result, no doubt, of the fact that most English port cities were run by an exclusive and competitive class of wealthy merchant families. These prominent entrepreneurs built their wealth on an ever-expanding maritime trade (particularly around wool) and also by working as customs officers and port authorities on behalf of the king. In addition to these legal trades, these families could also enrich themselves by supporting illicit piratical raids on competitors or ships carrying lucrative cargo from any nation.[10]

Throughout the Middle Ages, the English Crown and common law courts debated whether pirates were merely maritime robbers or if they were *hostis humani generi*—an ancient Roman legal category for pirates that insisted that they were traitorous "enemies of mankind" who challenged the very idea of governmental jurisdiction at sea. If they were merely seafaring thieves, pirates could be tried in common law courts where juries of their peers could decide if they were guilty or not. In many cases, juries were filled with men who profited from piracy, and pirates could be easily acquitted. On the other hand, if pirates were *hostis humani generi*, then the Crown could consider them seditious traitors and had jurisdiction to try them. Between the 1280s and 1350s, various English monarchs did try to reign in English pirates by establishing royal Admiralty courts and by ruling that independent maritime warfare was an act of treason against the royal prerogative. However, the lofty standards of evidence required to prove acts of treason in Crown-led Admiralty tribunals coupled with weak internal administration limited the effectiveness of this method of pirate hunting. Considering the widespread public support for pirates and sundry incoherent legal policies toward pirates, the English state was largely unable to stop its subjects from engaging in illicit maritime plundering throughout this era.[11]

To add to the legal confusion over what constituted piracy in medieval England, the English monarchs themselves oftentimes supported private naval raids during peacetime. Throughout medieval Europe, it was common for victims of maritime predation to apply to their monarchs for "letters of reprisal." These letters—often issued in times of peace—allowed the subjects of one nation to attack the shipping of the nation whose mariners had first attacked them and to seize cargo or goods equal to their initial losses. In essence, this was a maritime tradition equivalent to the ancient Babylonian principle of "an eye for an eye."[12]

While the reprisal system ideally would allow one's subjects to recover financial losses after an attack by pirates from another kingdom, it is easy to see how this system could also be beneficial to the Crown. After all, vessels operating under letters of reprisal would be required to give a share of their proceeds to the English king.[13] English monarchs could also utilize private naval power against their rivals. For instance, in the early thirteenth century, King John hired the French pirate and ex-clergyman known as Eustace the Monk to raid French king Philip II's shipping in the English Channel. Within a few years, Eustace betrayed his English paymaster and became a prominent French admiral.[14] This example both demonstrates the medieval English Crown's tolerance of private naval warfare when it benefited its strategic goals and the volatility such a service could provide.

Throughout the Middle Ages, the English Crown lacked any coherent semblance of a standing navy and depended upon private subjects and shipowners to make warfare at sea a possibility. The legacies of popular support for piracy in England, resistance to impressment policies by the Crown, the reprisal system, and Crown dependence on private naval warfare would all persist into the various corners of the nascent English empire of the sixteenth, seventeenth, and eighteenth centuries.

The English Crown, the Royal Navy, and Private Naval Warfare, c. 1500–1660

By the mid-fifteenth century, maritime trade in Europe transformed from a largely regional to an international affair. Much of this expansion was the result of changes in maritime technology. For instance, traditionally slow and hulky medieval cargo vessels such as the cog were supplanted by caravels and carracks—ships that balanced large cargo-carrying capacities with increased sailing power, range, and maneuverability. Portuguese and Spanish explorers used these new sorts of vessels to explore previously inaccessible regions in Africa and the New World and gained access to resources and wealth that gave them financial and military advantages over other European kingdoms.[15] The increased speed and sailing capacity of late medieval and early modern ships was augmented by a new and lethal tool: gunpowder. While the late medieval introduction of guns to ocean-going vessels did little to immediately change the face of war at sea, by the sixteenth century artillery would be a key factor in defeating or capturing enemy fleets.[16]

Aside from advances in naval technology, the way in which various Western European governments managed naval defense was also rapidly changing during this period. Historian Jan Glete argues that the gradual expansion of royal naval power was an important element of state building in Western Europe after 1500. By the late fifteenth century, the French, Spanish, and English states had all overcome various periods of civil strife and dynastic tension. Thus, these Western European monarchs were all simultaneously starting to "achieve a practical monopoly of violence in their territories." Glete contends that rulers of these nations—all of whom had significant Atlantic connections—started to create stronger laws regarding the monopoly of maritime violence, worked with private maritime interests to mobilize their resources and vessels for war, and created some semblance of more permanent naval forces.

Despite these impressive administrative changes, these fleets did not yet possess an entirely permanent structure. For instance, in the early sixteenth century, the Spanish Crown created the *Casa de Contratación*—an institution that granted trading and monopoly privileges to merchant interests in return for guarantees of service should the state need to assemble a fleet. Even though the Spanish Crown was getting more involved in matters of naval defense, its defense resources were still largely private in nature.[17]

Increased royal intervention in matters of maritime defense coupled with a continued reliance on private naval resources was also the order of the day in Tudor England. Beginning with the rule of the infamous King Henry VIII (r. 1509–47), the English Crown began to lay the groundwork for a more permanent royal fleet while also still depending on private naval resources to spearhead military expeditions. Henry VIII's split from the Roman Catholic Church coupled with ongoing military tensions with Scotland and France encouraged the monarch to construct over eighty-four new vessels to defend his beleaguered and increasingly isolated island kingdom. While these ships were built for the traditional medieval tactic of boarding one's foes, they were also equipped for the first time with scores of powerful guns for prolonged combat at sea.[18]

Henry VIII not only expanded the number of warships in his Navy Royal but more importantly created a permanent administrative network to oversee naval affairs. For the better maintenance of his growing fleet, the king ordered the construction of naval storehouses and docks at crucial port cities. Around the same time, a Council of the Marine formed that advised England's Lord Admiral and the king himself on naval affairs. While it is unclear who initially spearheaded the creation of this proto–Navy Board, the council included the new offices of the "Lieutenant of the Admiralty, the Treasurer, the Surveyor and Rigger of the Ships, and the Master of Naval Ordnance." While each of these officials managed various aspects of naval administration independently, they audited one another's accounts and provided a valuable advisory board for the Crown. For the first time in English history, the Crown not only controlled a fair number of warships for immediate military necessities but had an administrative network available for the management of future fleets.[19]

Even if Henry VIII's naval administration set the precedent for an increasingly Crown-controlled fighting force, he also recognized the importance of wielding private naval power. After all, Henry VIII broadened the medieval reprisal system—allowing his subjects to attack enemy shipping during times of war, whether they had been harmed by enemy raiders or not. This general governmental invitation to English subjects to wage private war on behalf of the state was an early form of what seventeenth-century Englishmen would later call privateering. Rodger explains that contemporary "language had not caught up with the new reality."[20]

It is important to note that the Tudor dynasty's support for private naval raiders occurred during a time period when English merchant vessels were getting larger, better armed, and more capable of sailing around the known world. In times of emergency or war, Crown officials could impress or hire these heavily armed ships as adjuncts to their own warships or could commission them to privately raid enemy shipping. Royal naval power during the early modern period, therefore, required a partnership between the Crown, the Crown's naval forces, and private naval actors.[21] The importance of this partnership between private and public naval interests was particularly noticeable during the reign of Queen Elizabeth I (r. 1558–1603), when the Protestant queen depended upon a combination of private naval warfare and the strength of the Royal Navy to harass Catholic Spanish shipping throughout the Old and New Worlds.

Not long after Christopher Columbus's fateful 1492 discovery of the New World, Pope Alexander VI decreed that Spain and Portugal were the sole powers allowed to trade, colonize, or even settle west of an arbitrary line drawn in the middle of the Atlantic Ocean. Spain—the predominant colonial power in the New World—took this papal mandate seriously and contended that any foreign interlopers were to be treated as enemies. By the mid-sixteenth century, Spain and its European rivals came to an understanding that peace treaties made in Europe would not necessarily apply in the Western Hemisphere—in other words, there would be "no peace beyond the line" for European competitors in the New World.[22]

By the late 1560s, a score of English merchants and slavers began to illegally trade with Spanish colonists in the New World. While French Huguenot traders and raiders had been the first foreign interlopers in the Spanish New World, Englishmen such as John Hawkins, Walter Raleigh, and Francis Drake quickly became the most iconic examples of English Sea Dogs who both illegally traded with Spanish colonists and led piratical raids on silver-laden Spanish ships and settlements. At one point, Raleigh even set up the first English colony in the New World—Roanoke (in modern day North Carolina)—with the hopes that it would become a maritime plundering base. Even though the settlement would prove to be short-lived, it was part of the increasingly provocative actions by royally aligned English adventurers with connections to Queen Elizabeth against the Spanish Empire. Rather than condemning peacetime attacks on her Spanish rivals,

Queen Elizabeth covertly encouraged Drake and his compatriots and even received a significant share of the Sea Dogs' illicit earnings.[23]

By the mid-1580s, the Crown's well-known covert support for the Sea Dogs coupled with rising religious and political tensions in Europe led to a global Anglo-Spanish War. Although the long conflict would end in a stalemate in the early seventeenth century, the Elizabethan conjunction of private and governmental naval interests provided the English with some notable successes. The most iconic case of this partnership came in 1588 during the failed invasion of England by the infamous Spanish Armada. Much of the Spanish defeat could be attributed to poor planning and foul weather. Nevertheless, Drake and other well-known Sea Dogs did successfully lead the combined forces of over thirty Royal Navy ships and nearly two hundred privately hired, levied, and impressed merchant vessels against the scattered Spanish squadrons that they encountered. While most of the poorly armed civilian ships remained out of the fight, some of the private vessels were well-armed ships on par with the warships of the Royal Navy and played a fundamental role in the routing of the Spanish fleet.[24]

The days when the queen's swashbucklers like Drake and Raleigh could lead relatively harmonious joint private-royal expeditions against the Spanish were numbered however. By the 1590s, even as the naval conflict with Spain continued, merchants increasingly began to shift their attentions away from privateering and toward increasingly profitable merchant trading missions in the Baltic, Mediterranean, East Asia, and eventually the Americas. After all, although trading vessels were undoubtedly armed, their commercial activities were often less risky than privateering ventures or serving as auxiliaries for the Royal Navy.

While civilian captains increasingly turned toward trading missions, they also began to complain more and more about impressment and the lack of protection from Royal Navy ships. Operational tensions also arose during this period between Royal Navy and private captains—particularly when it came to competing for mariners. The rift between the royal military and private mercantile entities became even more visible after the queen's death in 1603 and the accession of her distant cousin and the first Stuart dynasty king—James I. The new king ended his predecessor's long war with Spain but also inherited her naval debt. James's aversion to military spending and corruption within the naval administration of the Royal Navy led to a general decline in the quality of the government's fleet. The new king's novel unwillingness to tolerate private raids against the Spanish during peacetime led to a crackdown on would-be Sea Dogs with its most infamous result being the 1618 execution of Raleigh after leading an unauthorized assault on Spanish possessions in the New World.[25] All told, the Elizabethan paradigm of unity between public and private naval warfare against the Protestant kingdom's enemies had come to an end.

His aversion to private naval raids aside, it is important to note that King James did not entirely neglect his kingdom's naval defense. Throughout his twenty-two-year reign, the king did try to cut costs by limiting the size of the Royal Navy while also ordering the construction of several large warships including the fifty-five-gun *Prince Royal*. This 1,200-ton warship was emblematic of the king's desire for a small fleet of capital warships that would overawe international rivals through size and firepower—arguably an early modern display of what later centuries would call gunboat diplomacy. The king no doubt appreciated the fact that the construction of large warships would create a visible difference between his royal fleet and smaller privateering vessels. Nevertheless, these hulking ships were poorly equipped to hunt down the light galleys of the increasingly persistent North African Barbary corsairs and slavers that plied northern European waters for captives or the small craft of other groups of pirates. After James I's death in 1625, his son and heir, Charles I, would continue his father's problematic capital ship policy.[26]

By 1625, Charles I eschewed his father's attempts at peaceful diplomacy in Europe and planned for a new Anglo-Spanish war. He worked with his father's former Lord Admiral, the Duke of Buckingham, to launch an invasion of Cádiz, Spain, with an allied Dutch force. Ultimately, the ill-fated Cádiz Expedition would fail due to heavy parliamentary resistance to granting money to the king, a poor supply system, and the resistance of civilian ship captains to royal impressment. Sailors mutinied throughout English ports due to a lack of pay, and subsequent attempts to send joint royal-private flotillas against Catholic foes in Spain and France failed. Buckingham's control of the Royal Navy—complicated by all these issues—contributed to the ever-growing "gulf between the maritime community and the Stuart navy."[27]

This gulf came at a particularly troublesome time for the new king. Parliamentary resistance to Charles I throughout the disastrous fight against Spain and France brought an end to the new king's wars with those kingdoms. In addition to growing tensions with Parliament, Barbary corsairs continued to seize English merchantmen at the same time that naval forces from multiple continental European kingdoms engaged in the ongoing Thirty Years' War had started to threaten neutral English shipping. To enforce English claims of sovereignty in increasingly turbulent northern and western European waters, Charles—having already decided to rule the kingdom without a Parliament—reinstituted an old and extraparliamentary tax known as the "ship money" tax. Under this system, various counties would pay the Crown set rates for the construction of royal warships (including the hulking ship of the line *Sovereign of the Seas*). Historian Kenneth R. Andrews observes that even though Charles I's ship money fleet never saw much combat, regular peacetime deployment of an English Royal Navy—aided in no small part by the direct attention of the monarch himself—marked a "distinct step forward in the evolution of a professional state Navy out of the Elizabethan amalgam of royal and private enterprise."[28]

Whatever gains the ship money levies made for the future professional legacy of the Royal Navy, royal taxation without representation served only to amplify the growing political tensions between the Crown and advocates for greater parliamentary power during the 1630s. While Charles I may have dreamt of a centrally controlled Royal Navy that ruled the waves, some of his political opponents dreamt of resurrecting the age of Elizabethan private naval warfare. Among the greatest opponents of both ship money and the king himself were prominent Puritans such as Robert Rich, the Earl of Warwick. Warwick had a long history of private naval raids under both English and other European kingdoms at war with Spain. In the mind of Warwick and his freebooting allies, they could both serve God and harness significant profit by raiding Spanish shipping. Warwick's efforts culminated with his creation of the Providence Island Company—a joint stock company that used a small colony off the coast of Nicaragua to raid Spanish shipping and settlements throughout the 1630s.

Despite his boldness, Warwick's private naval war against Spain had absolutely no royal support. Nevertheless, Charles I—faced with increasingly violent resistance to his autocratic policies such as the ship money tax—could do little to stop the swashbuckling and freebooting of his most virulent critics.[29] By 1642, civil war broke out between Charles I, his Royalist supporters, and Parliament. While the king maintained some ships at his disposal, the majority of his officers and sailors supported the Parliamentary faction. Even though the reasons for this turn of events are not straightforward, it is likely that officers and seamen resented the king's attempts to install aristocratic "gentleman officers" in the top echelons of the Royal Navy as opposed to experienced mariners.[30]

Whatever led the formerly Royal Navy into Parliament's hands, the fleet was vital to securing the English Commonwealth's victory in European and American waters. While occasional English naval ships had ventured into the Western Hemisphere in previous decades, Parliament's naval forces—ironically commanded by the Earl of Warwick, the kingpin of English privateering in the Americas—would make the first major show of imperial naval force in the English Atlantic colonies in the early 1650s. Sir George Ayscue, along with seven warships, sailed to the New World with a twofold mission: to enforce the Navigation Acts—a new series of laws that forbade Anglo-American trade with English competitors such as the Dutch—and to subdue Royalist holdouts after the execution of King Charles I in 1649.

Broadly speaking, this show of imperial naval might have forced colonial submission to the Parliamentary regime with little if any armed resistance. This application of large-scale metropolitan naval action in the New World was a novel idea in the mid-seventeenth century and coincided with the Commonwealth's general use of the English Navy to wage war against the republic's Dutch and Spanish foes in European waters and beyond. It would also be during this interregnum period that the English government would begin to more regularly order state ships to convoy and protect commerce.

Notwithstanding the Commonwealth's expansive application of its navy, success could not last forever. Not long after Oliver Cromwell seized control of the Commonwealth in 1653, he devised a massive a thirty-eight-ship, three-thousand-man expedition to seize the important Spanish port of Santo Domingo on the Caribbean island of Hispaniola. This so-called Western Design failed miserably and ended in thousands of casualties from both battle and disease. Although his forces successfully captured the small Spanish island of Jamaica—no insignificant consolation prize—Cromwell's government was unable to continue to finance such costly campaigns. By the restoration of the Stuart dynasty and accession of Charles II in 1660, the English government had a strong central navy—in no small part shaped by the legacy of Cromwell and reformations of Samuel Pepys.[31]

Whatever the legacy of Cromwell's campaigns in the New World, Charles II's Royal Navy forces in the Americas were only moderately effective. To the English government's credit, English royal squadrons successfully captured New Amsterdam from the Dutch in 1664 and defeated a major French force at Martinique in 1667. Additionally, it was during Charles II's reign when Admiralty authorities started the policy of stationing guard vessels in stations such as Jamaica, Barbados, the Leeward Islands, and Virginia. Nevertheless, these stations were generally disease ridden and not popular with Royal Navy officers, and those officers regularly faced charges of corruption and mismanagement of their sailors while serving in the New World. Overall, the Royal Navy forces in the New World in the latter decades of the seventeenth century were ill-equipped to successfully guard entire colonies, to guard merchant vessels, or to hunt pirates.[32]

From the medieval era to the rule of Charles II, numerous rulers transformed the Royal Navy into an increasingly Crown-controlled and professional force. This professionalization had not fully crystallized as the first English colonies emerged in the New World. Financial and political restrictions kept the bulk of the still-growing Royal Navy in Europe, and Anglo-American colonists were largely forced to rely on familiar tactics ranging from emergency fleets to private naval raids to secure the safety of their infant coastal settlements.

Early Anglo-American Naval Defense Policies, c. 1630–1685

If the English navy was becoming an increasingly Crown-controlled institution by the mid-seventeenth century, the same could not be said for the largely autonomous and proprietary English colonies in the Americas. To understand the local nature of military defense in these colonies, one must first examine the constitution of the English Empire in the seventeenth century. Beginning with the founding of Jamestown, Virginia, in 1607, English colonists in the hundreds, thousands, and tens of thousands established wide-ranging colonies from Newfoundland, Canada, to Barbados in the West Indies. By the 1640s, around fifty thousand colonists—primarily of English descent—had settled in the small English strongholds between Canada and the West Indies. The New England settlements from Maine down to Rhode Island may have been among the largest population centers in this era but still collectively had fewer than twenty thousand settlers by this period.[33]

Although Anglo-American colonists were subjects of the English monarch, the Crown could not rule its colonies by fiat. In the decades following the establishment of the first permanent English colony at Jamestown, Virginia, in 1607, Anglo-American colonists developed representative bodies and legislatures that oversaw the day-to-day operations of these settlements. Despite rare shows of force such as the English Republican navy's campaign against Royalist holdouts in the colonies in the 1650s, the English government was largely institutionally and militarily unable to control its dominions in the Americas for much of the seventeenth and eighteenth centuries.[34]

If the Crown did not have the institutional capacity to enforce its will on Anglo-Americans in the seventeenth century, it certainly did not have the military capacity to meet their every defensive need. With this in mind, the Stuart monarchs empowered colonial authorities to manage their own defenses with little guidance from London. This laissez-faire approach to colonial military governance, a policy that would guide royal-provincial relationships until the late 1740s, was evident at the highest level of colonial governance through royal instructions to colonial governors. For instance, in 1628/9, King Charles I commissioned the Massachusetts Bay Colony to "at all Tymes hereafter for their speciall Defence and Safety, to . . . resist by Force of Armes, as well by Sea as by Lande, and by all fitting Waies and Meanes whatsoever, all such . . . Persons, as shall at any Tyme hereafter, attempt or enterprise the Destruccon, Invasion, Detriment, or Annoyaunce to the said Plantation or Inhabitants, and to take and surprise by all Waies and Meanes whatsoever, all and every such Person and Persons, with their Shippes, Armour, Municons and other Goodes."[35]

With such open instructions to manage military defense however they thought necessary, it is unsurprising that early Massachusetts authorities adapted the well-established English military traditions that had been so familiar to them in their homeland for use in the New World. Early on, Massachusetts authorities established a provincial militia system that continued some English traditions (e.g., the requirement of all adult men to serve in the militia) but adapted their militia system to reflect provincial values (e.g., the election of officers). The colony's military laws were codified in 1647 in *The Book of General Laws and Liberties*.[36] While this early American legislative record demonstrates the continuity and changes in various aspects of English militia laws, it also demonstrates New England authorities' willingness to continue the ancient practice of impressment of vessels for military purposes. This is clear in the provision that "it is intended, that the general words aforementioned, contain in them power to impress and send forth soldiers[,] . . . vessels at sea, carriages and all other necessaries, and to send warrants to the treasurer to pay for the same."[37]

These measures would be particularly important due to the precarious military situation the colony faced in its first decade of existence. Beginning in the 1630s, New Englanders found themselves in a "multidirectional struggle for control of the coast" with Dutch and Algonquian-speaking Indigenous rivals.[38] The importance of Anglo-American provincial naval power would first become obvious in the Pequot War of 1636–37. This largely maritime conflict began in 1636 amid a confusing and bloody matrix of cultural misunderstandings and imperial rivalries. While peaceful trade did occur between the various groups in the Northeast, murders, raids, and counterraids between the Pequots, Narragansetts, Dutch, and English mutually escalated tensions throughout the region. In 1636—citing the killing of two English traders at sea by Pequots and Narragansett-aligned Manisses—Massachusetts Puritans spearheaded an invasion of Pequot-held territories on Long Island Sound and the Connecticut River valley. Attacking the Pequots as opposed to other tribes was a strategic decision; the "friendless" Pequots had alienated most other Native sachems in the region and controlled a large swath of important maritime territory that limited Anglo-American expansion in the area. In the minds of Anglo-American authorities, eliminating this Indigenous threat would open up vital trade routes south of Boston.[39]

Broadly speaking, the various small New England settlements involved in the conflict (ranging from Massachusetts to the future colony of Connecticut) did not have the capability to build or acquire substantial naval forces to fight the Pequots at sea. Historian Howard Chapin was correct that armed English merchants engaged in early battles with the Pequot "can scarcely be classed either as privateersmen or colony coast guards."[40] Many of the maritime battles in the Pequot War involved small-scale shootouts between crews of Anglo-American shallops and pinnaces against Pequots on canoes, pinnaces, and shore parties. Despite having gunpowder and artillery at their disposal, Anglo-American authorities and their mariners were frequently outpaced and outsailed by Pequot war-canoe crews who mastered guerilla tactics. Ironically, for a war that began at sea, it would be the land-based Anglo-American massacre of Pequot residents of Mystic Fort in the summer of 1637 that secured victory for the New England combatants and their Indigenous allies among the Narragansetts and Mohegans.[41]

Despite the small scale of naval combat in the conflict, the impressment and arming of small craft for transports and gunboats to fight Pequot war-canoe crews would set an important precedent for how the various New England colonies would wage war at sea against Indigenous and European foes for the next century. An illustrative example occurred in May 1637 when the newly formed Connecticut General Court impressed a shallop from trader John Pynchon for a riverine expedition against the Pequots. This shallop would be one of three vessels including a pinnace and pink that ferried over ninety Connecticut and allied Indigenous soldiers. Pynchon would later protest the impressment of the vessel, particularly after the fledgling Connecticut government at Saybrook also passed heavy taxes on his village of Springfield for support of local forces during the conflict.⁴² As will be recalled, governmental impressment of merchant vessels for naval expeditions coupled with social resistance to the practice had been common occurrences in England for centuries.

It was during this early era of English colonization that we also get a glimpse of how Anglo-American officials financed these early provincial naval organizations. While provincial authorities could forcibly impress vessels in emergencies, this strategy would not be feasible without some sort of financial compensation for those affected shipowners. For instance, when a mysterious ship sailed close to Boston Harbor in the spring of 1645, the colony's legislature—the General Court—ordered one Major Edward Gibbons to "send two shallopps furnished with men" to investigate "what the shipp that lyeth [hovering] about these coasts is . . . the chardges to be defrayed out of the custome of wyne."⁴³

While most provincial naval organizations in seventeenth-century New England were temporary emergency fleets, there were some early strides made toward keeping a semipermanent coast guard during this era. In the early 1640s, ongoing Dutch and Native threats coupled with the concurrent disruptions caused by the English Civil Wars encouraged four New England colonies—Massachusetts, Plymouth, New Haven, and Connecticut—to form a novel military alliance known as the United Colonies of New England or the New England Confederation. One of the primary foes of the confederation would be Ninigret, a sachem of the Narragansett people who furiously opposed Anglo-American expansion in the Northeast.⁴⁴ Commissioners from all four colonies commissioned one Captain John Younge to take a vessel and up to twelve armed men to cruise throughout the Long Island Sound. His orders were to reassure friendly tribes such as the Montauk while also countering Narragansett moves at sea.

Far from considering this a temporary mission, the commissioners created a provision for Younge's service to continue to the spring of 1656 or beyond. The New England Confederation's minutes from September 1656 indicate that the allied colonies jointly paid over £153 for the hiring of Younge, the crews of his bark and shallop, eight soldiers that likely acted as marines, and various provisions including gunpowder. Intriguingly, some of the cost of the expedition was covered by tribute paid by the recently defeated Pequot people.⁴⁵ This early example of a joint colonial effort to finance a provincial guard vessel foreshadowed the more organized and semipermanent nature of provincial navies of the next century.

Other fledgling colonies to the south of New England also adopted local measures for coastal defense—even when the rare Royal Navy force was present or patrolling in the area. One dramatic example of an early provincial emergency fleet occurred in Virginia in 1667. Virginia—a vital tobacco-producing province—had become the first royal colony on the North American mainland with a Crown-appointed governor in the 1620s.[46] After Virginia's provincial government pleaded for commerce protection by the Royal Navy in the early months of 1667, the Lord High Admiral in London dispatched the forty-six-gun frigate *Elizabeth* to patrol the Chesapeake Bay. This vessel would be the first Royal Navy station ship in mainland British North America. While HMS *Elizabeth's* arrival in 1667 may have temporarily assuaged provincial anxieties, it was so badly damaged from its Atlantic crossing that it would take months to repair. Shortly thereafter, a squadron of five Dutch warships entered the Chesapeake and quickly captured the outmatched Royal Navy vessel. Without even the mirage of protection from a Royal Navy vessel, Governor William Berkeley quickly assembled the colony's militia and attempted to impress various merchant vessels for an emergency fleet.[47]

Records of the negotiations between Berkeley and the ship captains not only demonstrate the complex politics behind impressing and creating provincial emergency fleets but also the continuities of older ad hoc naval defense strategies from the Old World. In a letter to the Earl of Arlington, Virginia councilman Thomas Ludwell described the reluctance of ships masters on the York River to voluntarily resist the Dutch invasion fleet. They insisted that they would anger their merchants "and owners if they voluntarily brought theire shipps & goods into hazard, and therefore desired they might be pressed into the Kings service and have security given them for all damages they might receave from the enemy." In other words, these merchant captains suggested that an official act of impressment would relieve them of responsibility for damages incurred in battle with the Dutch. Although it is unclear who the shipowners were, it is highly likely that they were English merchants engaged in the lucrative importation of tobacco back to London.

With the concerns of the shipmasters in mind, Berkeley ordered the king's "Broad Arrow"—a symbol of royally seized military stores—carved on the masts of nine vessels to show visible evidence that these vessels had been impressed into military service. Berkeley had the vessels' value appraised and promised that both he—and even King Charles II himself—would cover the damages the impressed vessels would incur, promised compensation for wounded sailors, and promised that the merchant crews could keep all plunder. According to Ludwell, Berkeley then prepared to lead a fleet of nearly "400 men and boys . . . many of them sick" under his command. This force included the captain and crew of the recently captured Royal Navy ship *Elizabeth*—against the far inferior Dutch force. Despite this impressive show of force, Ludwell alleged that "cowardly feare" on the part of the merchant captains delayed the fleet so long that the Dutch were able to flee the Chesapeake with several prizes.[48]

Whether or not the Virginian's charges of cowardice were true, this episode reveals several important features of provincial and Royal Naval defense strategies in the second half of the seventeenth century. First, the ancient strategy of impressing private vessels for emergencies was familiar enough throughout the English Atlantic world for merchant captains to request an official act of impressment to shield them from criticism from shipowners in London. Second, the sole Royal Navy ship stationed on the North American mainland was so inoperable that a royal governor had to create his own defensive fleet to secure his shoreline. In a microcosm, the provincial response to the Dutch invasion of Virginia demonstrates the limits of royal military power in the colonies and the flexible defense options available to provincial governors in the absence of royal military aid.

In England's lucrative West Indian colonies, provincial governments also relied on local naval power in the wake of limited royal assistance. Like their mainland compatriots, Caribbean governments directly fit out provincial guard vessels when necessary. Take for instance the Barbados Assembly's 1667 plans for the hiring and fitting out of a provincial navy during the latter years of the Anglo-Dutch War. They called for "two of the nimblest ships of force [to] be victualled, manned, and armed for guarding the coasts of the island" and impressed provisions, guns, and boats for 180 sailors and the two vessels for a month's service.[49]

While Caribbean authorities throughout the second half of the seventeenth century directly established temporary provincial navies, they also relied heavily on the service of "private men of war" or "privateers." Intriguingly, the very term "privateer" first appeared in English records in the 1660s to describe the thousands of English sailors—particularly in Jamaica—who made their living by seizing Spanish prizes with commissions from sympathetic provincial authorities during the interregnum. Even though Charles II's accession to the throne in 1660 meant that Cromwell's former war against the Spanish was over in Europe, Jamaica's first royal governor—Thomas Hickman Windsor, or Lord Windsor—was convinced that there was "no peace beyond the line" and continued to issue licenses to buccaneers like the infamous Henry Morgan to raid Spanish shipping. In much the same manner as Queen Elizabeth in the previous century, Windsor supplemented his income by taking shares from these peacetime private naval raiders.[50]

While there can be no doubt that there were financial incentives for provincial governors to encourage privateering against their nominal foes, there were also potential strategic benefits to such a practice. Although the governors never had express royal instructions to use Royal Navy vessels or privateers to attack Spanish targets, Charles II ordered Jamaican officials to convince their Spanish neighbors to trade—even with force if necessary. With such orders in mind, Windsor and his successor Sir Charles Lyttleton commissioned Sir Christopher Myngs—a noted buccaneer—to lead joint Royal Navy–privateer raids on Spanish targets throughout the Caribbean.[51]

The fusion of royal and privateering resources in these expeditions speaks to the still-weak nature of the Royal Navy in the West Indies in the late seventeenth century. While there had been cases of sizeable Royal Navy squadrons operating in the Americas during the intermittent conflicts with the French and Dutch in the 1660s and 1670s, these fleets rarely stayed in the New World for more than a few months at a time. Those few royal station ships throughout the New World—always numbering under a dozen—could do little to enforce the Crown's prerogative on the seas or to protect English commerce. The power dynamic between privateers and the Royal Navy is best exemplified by Morgan's 1668 seizure of the HMS *Oxford*—a Royal Navy vessel that had been sent to restrain the exceedingly violent wave of Caribbean buccaneers.[52]

Between the sixteenth and seventeenth centuries, the English Royal Navy grew into an increasingly professional, complex, and centrally controlled force. Nevertheless, the Crown only made piecemeal attempts to use this slowly evolving force to its advantage in the New World. To defend their shores and commerce, officials in the infant English colonies on the North American mainland and in the West Indies drew on old maritime strategies ranging from the impressment of merchant vessels to the commissioning of privateers. Anglo-American reliance on both of these strategies would only increase by the end of the century as England—and its global empire—would become enmeshed in a generation of world wars with France and Spain.

2

Provincial Navies in the First Imperial Wars of 1689–1713

In 1688, English rebels dethroned the unpopular Catholic and absolutist King James II and installed the Dutch Protestant William of Orange and English Mary as the empire's new monarchs. Shortly thereafter, Anglo-Americans initiated similar uprisings against the former king's officials throughout several colonies. This religious and political revolution on both sides of the Atlantic not only transformed England's Atlantic political makeup but triggered nearly two and a half decades of imperial conflict with France.[1] Anglo-American provincial leaders from Canada to Barbados had some previous military experience but were woefully unprepared for the global conflicts with England's imperial enemies known as the King William's War (1689–98) and Queen Anne's War (1702–13).

Historians have long recognized that these lengthy and expensive conflicts forced colonial governors to rely on long-standing civilian militias in lieu of the small number of red-coated royal troops in America.[2] Historian Don Higginbotham has observed that by the eighteenth century, Anglo-American infantry forces "had advanced from seventeenth-century militia to . . . eighteenth-century semiprofessional forces" in the decades preceding the American Revolution.[3] Arguably, a similar "military evolution" is evident in colonial Anglo-American maritime defenses during the imperial clashes of 1689–1713 as well. Colonial participation in what were essentially world wars would require greater financial investment in provincial military organizations, greater social mobilization for longer conflicts, and expanded vigilance in patrolling vital waterways for enemy privateers and invaders. English colonies that had previously mobilized small provincial fleets would expand upon these institutions at the beginning of the eighteenth century. In both King William's War and Queen Anne's War, provincial officials would plan major naval assaults on enemy ports and would also create semipermanent defensive fleets enshrined by local laws and funded by local resources. Although these provincial navies were impressive in scope, controversies surrounding their creation and maintenance amplified sociopolitical tensions within colonies and raised larger questions about the role of the Crown in defending its American provinces.

The Expanding Roles of Provincial Navies in the Atlantic World, c. 1689–1713

By the 1680s, the English Empire in the New World was expanding. Older colonies such as Massachusetts and Virginia were soon joined by Pennsylvania and the Carolinas. Nevertheless, with the onset of King William's War in 1689, the few Royal Navy vessels stationed in an increasingly expansive English Empire in North America and the Caribbean were ill-equipped to wage war against the French. These smaller frigates—usually tasked with patrolling for pirates and smugglers or with convoying English merchant vessels across the Atlantic—included fourth rate frigates that had fifty or more guns, fifth rates that had between thirty and forty-eight guns, and sixth rates that had between twenty and thirty guns aboard. Only on rare occasions did larger third-rate warships (with sixty to eighty guns) enter the North American or West Indian theaters.[4]

Even though the occasional guard vessel plied North American waters, imperial naval planners always devoted more naval attention to the West Indies than to North America—a trend that would continue throughout much of the next century. For instance, in 1701, while there were only a few Royal Navy station ships on the North American coastline, there were nine Royal Navy ships stationed at Jamaica alone in addition to the much larger West Indian fleet led by Admiral John Benbow. There were a number of reasons for this uneven distribution of Royal Navy assets throughout the New World, including the colder conditions in mainland North America, the higher revenue of the plantation islands, the necessity of threatening Spanish pretensions in the heart of its New World empire, and the British government's yearning for control over the Spanish bullion trade in the Caribbean.[5] This imperial bias toward the defense of the West Indies became more pronounced as Queen Anne's War progressed with expanded numbers of guard vessels, victualling stations, and maintenance facilities.[6] Even though the seeds of greater Royal Naval involvement were being sown in the Caribbean during the first decades of the century, the full "royalization" of naval warfare in the Atlantic world would take another half century.

Even with the presence of a Royal Navy guard ship in port, tranquility and coastal security were never guaranteed. This was evident in 1686, when Admiralty authorities sent Captain John George with the small frigate HMS *Rose* to guard Boston in 1686 after King James II's establishment of the Dominion of New England. With enterprising administrators such as Edward Randolph and Governor-General Edmund Andros at the helm of the Dominion, James II hoped to consolidate royal power by combining the administrations of every colony between Massachusetts and New Jersey.[7] Much to the chagrin of the Dominion's critics, Captain George allegedly used his position for profiteering, served as a yes-man to Andros and Massachusetts official Joseph Dudley, and did little to pursue the region's increasing number of pirates.[8]

During the late seventeenth century, Royal Navy recalcitrance in pirate hunting coincided with widespread provincial support for piracy against Spanish trade. As early as the 1670s when rulers in West Indian colonies such as Jamaica and Barbados began to shun those pirates that had orchestrated peacetime raids on Spanish targets, those enterprising Anglo-American pirates sought out new markets in North American ports. These swashbucklers found willing customers among proprietary and charter colonies with looser royal governance, including Massachusetts, Rhode Island, and Carolina. Royal centralizers, including Andros's political ally and administrator Randolph, began to equate this trending support for piracy with a worrisome provincial desire for political autonomy.[9]

While some New Englanders may have seen trading with pirates as a way to subvert royal trade laws, they mainly welcomed Spanish bullion-bearing pirates for economic reasons, particularly during the postwar economic depression that followed the destructive King Philip's War of the mid-1670s. During this conflict, Massachusetts authorities even employed a few former West Indian buccaneers familiar with guerilla combat to pursue Indigenous forces.

The region's flirtation with piracy would soon face significant resistance however. With increasing Spanish pressure on James II by the late 1680s, the monarch ordered Andros and Randolph to crack down on piracy.[10] These royal instructions encouraged Andros to utilize both HMS *Rose* and the temporary station ship HMS *Kingfisher* to crack down on pirates and English merchants violating the Navigation Acts.[11]

With such few Royal Navy ships at their disposal, it is unsurprising that Dominion authorities hoped to augment those guard vessels with provincial naval forces. This utilization of provincial warships to supplement Royal Navy patrols and cruises would ultimately become commonplace throughout the English Atlantic world. On May 25, 1687, Randolph suggested that "itt is necessary [that] a Small vessel be provided for his Majesties Service On the Coasts."[12] Randolph's suggestion ultimately led to the provincial government's purchase of the *Speedwell* ketch.[13] Throughout the next several years, provincial authorities would use the *Speedwell* for many tasks that would become routine for Massachusetts provincial navy vessels: ferrying soldiers, transporting supplies, and even escorting high-ranking officials to the contested Maine borderlands and Canada.[14] While provincial governments could occasionally hire privateers to conduct such missions, military vessels under their immediate control and supervision would prove to be more reliable for immediate strategic needs.

By early 1688, Anglo-American tensions with the Franco-Indigenous forces on the Maine borderlands convinced Andros to increase the English military presence there, including one of the region's first multivessel flotillas.[15] In 1690, Andros reported that throughout 1688–89, "The severall Vessells Imployed for the security of the Coast and fishery of that time were His Majesty's Sloope *Mary* John Alden Comander . . . His [Majesty's] New Sloope *Speedwell* John Cooke Comander, finished and ready to take in stores and provisions for the Eastward," the sloop *Sarah*, and the brigantine *Samuel*.[16] It is worth noting that Andros neglected to mention at least one other provincial vessel that was in service, the sloop *Resolution*.[17] Comprised of small coasting vessels such as sloops and brigantines, Andros's navy was clearly designed for coastal patrols and reconnaissance rather than pitched naval battles with enemy forces. With the expansion of this fleet, Andros had arguably created the first semipermanent provincial navy that the English colonies had yet seen.

Despite its utility, however, international and local political controversies would soon end the operations of this fleet almost as soon as they had commenced. Many New Englanders had begun to grow weary of Andros's strict military discipline on the Maine frontier and also fumed over rumors that the Catholic King James II had a new child and heir.[18] These developments, combined with Andros's Anglican leanings and his widespread eradication of provincial legal autonomy, were too much for Puritan-leaning Bostonians. With news that the Dutch Protestant William of Orange had landed in England and dethroned King James II, Protestant rebels led bloodless revolts against James II's officials throughout many of the American colonies. In April 1689, provincial authorities led over two thousand militiamen in a coup against Andros and imprisoned him and other Dominion officials before sending them to England.[19]

Outside of this political revolt, *Rose* and Andros's provincial navy also attracted the ire of the rioters. Deserters from *Rose* reported that the unpopular Captain George—with a Catholic lieutenant under him—planned to attack Boston with Andros and hand Boston over to the French. Those absconders, coupled with an angry mob of Boston rebels, dismasted the ship while it was docked in the harbor.[20] Beyond this attack, one shipbuilder even complained that rioters took the sails off an unnamed sloop that he had built for the Andros regime.[21] While this sloop's name was never mentioned, it is possible that this vessel was *Mary*. With widespread contemporary rumors that one of Andros's soldiers (an alleged Roman Catholic) planned to seize *Mary*, it seems that New England authorities had equal reason to fear that their "Papist" enemies would use royal vessels and their own provincial ships against them.[22]

Whatever the political motives of the New England rebels, the revolutionary government in Boston—with Simon Bradstreet as its new governor—would soon have its hands full with fallout from this uprising, an outbreak of piracy, and the beginning of an imperial war with France and its Indigenous allies. While they dissolved the crews of *Rose* and the rest of Andros's provincial navy, Massachusetts authorities quickly reemployed some of the colony's sloops to meet these threats.

In fact, outside of reforming some of Andros's provincial navy, the Massachusetts government also added new and larger vessels to the region's ever-growing list of provincial guard ships. This was particularly evident in the mid-1690s, when the provincial government—concerned that Royal Navy ships would be useless in shallow shoal waters off the coast—commissioned the *Province Galley*. The *Province Galley* was a two-masted, ten-gun warship that had oars to propel it through shallow waters and to pursue enemy craft. As it would happen, there would be two such *Province Galleys* throughout the rest of King William's War and Queen Anne's War.[23]

As scholars of naval history well know, navies will always have their detractors. One contemporary critic complained that the initial *Province Galley* was a "small vessell about 70 Tuns" that would "not be able to make any considerable Defence if [a number of enemies] should board" her. The critic did contend that *Province Galley* would "do sirvice upon some small priviters."[24]

Despite this skepticism, *Province Galley* would prove to be a major addition to Massachusetts's maritime defense capabilities. By the end of King William's War, Massachusetts governor Stoughton was able to brag that *Province Galley*'s Captain Cyprian Southack and his crew were "constantly employed to cruise about the Capes and convoy vessels from Virginia, Pennsylvania, Connecticut, etc. between Massachusetts, Martha's Vineyard, and Rhode Island. She has been of great service and the Commander has acquitted himself with great care and diligence, none of the vessels under his charge having miscarried." *Province Galley* not only served as a provincial guard ship for Massachusetts but as a regional guardian for English commerce throughout the northern Atlantic. This was a clear step beyond the impressment of local merchant ships for temporary service—an evolutionary process that began, as we saw in the last chapter, with the New England Confederation's hiring of temporary guard vessels during the fight with Ninigret in the 1650s.[25]

While imperial warfare saw an expansion of provincial naval operations in the northern colonies, West Indian governments also began to fit out local defense fleets on a larger scale than ever before. Early twentieth-century historian Ruth Bourne argued that Anglo-American governments in the Caribbean during the Queen Anne's War were "helpless and open to the enemy, unwilling and almost unable to cooperate with each other." For Bourne, neither "local sloops, merchantmen, privateers, nor convoys adequately reenforced the few [Royal Navy] cruisers" in the West Indies.[26] Despite this broad assertion, both primary evidence and subsequent scholars have emphasized the importance of provincial naval vessels in the West Indian theater during King William's and Queen Anne's Wars.[27] The importance of these vessels was heightened by the fact that the region's Royal Navy forces struggled with tropical disease and a haphazard provisioning system.[28] Depletion of Royal Navy manpower and supplies meant fewer Royal Navy warships could defend the island colonies—thus necessitating the presence of more provincial guard vessels.

With hostile French or Spanish forces often only a few islands away, West Indian officials frequently went to great expense to shore up coastal defenses while waiting for that much-desired Royal Navy assistance. In a 1689 letter to the Lords of the Committee for Trade and Foreign Plantations (later known as the Board of Trade), the Leeward Islands' governor Christopher Codrington lamented that his government was forced to levy a heavy "Tax of one million of Sugar" to supply infantry units and provincial naval vessels but also bragged that a "Privateer and my own two Sloopes are arrived here with a French Briganteen and two French Sloopes."²⁹ Codrington's letter reveals not only one common source of funding for provincial guard ships—local commerce taxes—but also the vague distinction between a governor's own warships and "privateers." The malleability of terms describing government-commissioned warships became so vague by Queen Anne's War that one of Codrington's successors—Governor Daniel Parke—would confusingly refer to a government-commissioned private raider as a "publick privateer."³⁰

Even though categorical haziness between "privateers" and "provincial navies" would persist throughout much of the next century, most provincial officials seemed to draw a distinction between government-operated naval forces and privately commissioned warships. One of the most dramatic instances of this distinction came at the beginning of Queen Anne's War in 1702 when Barbadian authorities complained about the "inconveniences of granting Commissions to privateers at this time, for that the vessels taken up for the service of this Island and defending our coasts do want sailors" and decided to prevent privateer ships from sailing while officials fitted out provincial "vessels of war."³¹ While privateers were useful for raiding enemy commerce, local governments often preferred to have at least some vessels under their immediate command during emergencies and sustained military expeditions.

While colonies throughout the English Atlantic built provincial navies to guard their commerce from enemy raiders whether or not Royal Navy forces were nearby, two consecutive decade-long conflicts also saw the frequent utilization of provincial warships to spearhead expeditions against enemy ports and to support infantry operations on land. One clear example of provincial naval support of infantry operations occurred in the spring of 1703/4, when Massachusetts governor Joseph Dudley expanded his colony's militia and naval forces and ordered the colonial ranger Colonel Benjamin Church to assault French-aligned Wabanaki forces on the Maine borderlands.

Dudley and his partners in the provincial assembly cooperated to establish a provincial naval pay scale titled an "Establishment of the Pay for Vessels Taken up for War & Transports & Officers & Mariners Pay." Aside from *Province Galley* itself, the Massachusetts government's financial establishment allowed the colony to fit out several other armed vessels and dozens of transports.³² During the expedition, Church himself convinced the Royal Navy captains on his expedition that it was "very expedient and serviceable to the crown, that Captain Southack in the [*Province Galley*] should accompany them [on a patrol], which they did readily acquiesce with him in."[33] Far from resenting the presence of provincial vessels on campaigns, Royal Navy officers came to depend on them for vital assistance.

While provincial warships were useful for offensive naval patrols and offensive campaigns, colonial governments also used them for diplomacy with various coastal Indigenous nations. For instance, in the spring of 1701 as war seemed more and more likely with France and Spain, the Massachusetts governor and council made various military preparations including reinforcing Castle William in Boston Harbor and *Province Galley*. The officers and crew of *Province Galley* were then ordered to escort government commissioners to "Casco bay, there to meet with and discourse the Eastern Indians; and to endeavour to hold them Steady to his Majesty's Interests and That the value of One hundred Pounds be sent by them for Presents."[34]

In addition to "soft power" missions to Indigenous neighbors, New England's provincial naval forces were also used for military strikes on the colony's Indigenous enemies. For instance, in May 1705, the *Boston Newsletter* newspaper reported that Governor Dudley sent *Province Galley* and another local vessel to pursue "5 or 6 Canoo's of Indians" that had attacked an English fishing shallop near Winter Harbor, Maine.[35] In essence, provincial navies were both useful for diplomatic and punitive expeditions with Massachusetts's Indigenous allies and enemies.

Each of those aforementioned naval missions (including guarding coastal commerce, attending to land-based infantry operations, and even pirate hunting) did not occur in a vacuum and frequently all occurred simultaneously. For instance, even as Dudley's administration planned a major offensive against the Wabanaki, he also outfitted small sloops to guard merchant vessels between late 1703 and early 1704 and commissioned other vessels to hunt down infamous privateer-turned-pirate John Quelch.[36]

This ability to wage three different provincial naval campaigns and patrols within the space of a few months points to a growing provincial commitment to naval warfare as long imperial wars dragged on. Yet these expanded naval operations did not come without economic and political costs. For instance, in July 1704, Dudley bragged that the General Court had "very frankly granted [£23,000]" to the fitting out of Church's naval and land expedition. Nevertheless, he openly lamented that citizens within Massachusetts were "oppressed with hard marches and great taxes" while their neighboring colonies did not share the burden.[37] As we shall soon see, concerns over taxes and insufficient naval assistance from other colonies were among the many larger sociopolitical and economic costs of provincial naval warfare.

Even though provincial navies were most commonly devoted to the immediate security needs of their home regions, the most consequential and controversial deployment of provincial naval forces were the rare occasions when they were committed to larger assaults on distant enemy strongholds. While colonial vessels had taken part in some assaults on enemy settlements during the seventeenth century, it was truly during King William's and Queen Anne's Wars when Anglo-American provincial navies first took part in major assaults on French and Spanish port cities.

In victory or defeat, provincial naval and land assaults on cities were always costly in terms of money, shipping, munitions, and most importantly manpower. The Massachusetts government's painless capture of the French Nova Scotian base at Port Royal (later Annapolis Royal) in early 1690 was uncharacteristic as far as those expeditions went. In the late spring of 1690, Sir William Phips led a force of more than seven hundred men on five vessels— including the provincial navy vessels *Six Friends*, *Porcupine*, and *Mary*—to capture the French port. After the force arrived on May 9, the small French garrison surrendered the town and ramshackle fort without firing a single shot. Phips's men sacked the town and enjoyed the simple victory.[38]

While Port Royal had been relatively easy to capture, the subsequent New England assault on Quebec later that summer would be an utter failure. Without Royal Navy ships at their disposal, provincial authorities from Massachusetts, Connecticut, and New York planned a joint land assault on Montreal, and Massachusetts led a naval assault on Quebec.[39] While this autonomous intercolonial military alliance may seem noteworthy, it was mostly done out of convenience and necessity during a period when royal military assistance was unavailable. Far from preferring such independent military measures, most colonial governments desired significant royal military assistance.[40] For instance, as early as March 1690, Bostonian Elisha Hutchinson opined that "If his majesty would please Speedily to furnish us with two ffrigatts and Amunition" the taking of French Canada would be possible.[41]

Despite the absence of Royal Navy ships, the expedition proceeded as planned. It is interesting that the editor of the first issue of the first newspaper ever published in English North America, the short-lived *Publick Occurrences* of Boston, boasted that Massachusetts native Phips commanded a "Navy of two and thirty Sail; which went from hence the beginning of the last August" against Quebec.[42] Phips's massive colonial navy was only possible thanks to widespread impressment of private vessels and a large loan from Boston merchants. As the colony's government would soon discover, even these emergency measures were not adequate for the expedition's 2,300 soldiers and sailors.[43] Over the next few months, disaster stalked the ill-prepared provincial forces. While the land assault on Montreal never materialized, the maritime attack on Quebec failed due to late autumn storms, inadequate supplies, disease outbreaks, and shipwrecks. Ultimately, more than four hundred New Englanders would lose their lives throughout the catastrophic and aimless campaign.[44]

Provincial authorities were not only overwhelmed by the protests of angry mariners and soldiers but also by a hefty expedition-related debt. Connecticut issued the first major taxes since Andros's rule, and the colony of Plymouth—which was soon to be subsumed by Massachusetts—raised tax rates to higher levels than had ever been demanded. Massachusetts's debt—compounded no doubt by the loans from Boston's merchants—rose to nearly £40,000. In response, the government in Boston took the controversial step of issuing paper bills of credit to stimulate the shattered economy.[45] One merchant, worried over the fervor of unpaid sailors and soldiers from the expedition, claimed "we have found a way to stop the mouths & aswage the passion of the: soldiers & seamen by a new mint raised here of paper money . . . there are not many that take it & they that have it scarce know now what to do with it."[46] Despite these criticisms, scholars have long recognized that Massachusetts's novel adoption of paper money—partly inspired by provincial naval costs—set a standard for many other colonies to adopt paper currency to pay for immediate war-time measures throughout King William's War and Queen Anne's War.[47]

On the other side of the nascent English Empire, provincial naval power also played a fundamental role in South Carolina's first (and ill-fated) expedition against Spanish Florida. As early as 1701—a year before the outbreak of Queen Anne's War—South Carolina's government hired a small number of Anglo-American and Indigenous mariners to operate a fleet of scout canoes and to look out for Spanish incursions on the colony's coast. The next year, with unconfirmed reports that England had gone to war with Spain, Governor James Moore planned a general invasion of the Spanish port city of St. Augustine. As South Carolina was a proprietary rather than a royal colony, Moore could expect little direct assistance from the Crown's military forces. Like his New England compatriots in the 1690 expedition against Quebec, Moore impressed and armed at least ten merchant ships—and used several dozen more for troop transports for the colony's militia.

While Moore's preparations were grandiose for the young colony, he had doubts as to whether or not provincial forces alone could take St. Augustine. In a letter to the governor of Jamaica, he expressed worries that French soldiers were helping the Spanish to man the city's large fortress—the Castillo de San Marcos. He intimated that if that were the case, it would "never be subdued but by a royal force and Navy, which will be an extraordinary but necessary charge to the Crown." Although French forces were not actually present as he feared, fierce local resistance and the arrival of Spanish warships forced the Carolinians to burn their ships and flee back to their capital at Charles Town.[48]

By the latter years of Queen Anne's War, Royal Navy vessels began to accompany provincial forces on these expeditions in greater numbers. In 1707, Massachusetts provincial and Royal Navy forces attempted to capture French Port Royal (which had reverted back to French control during King William's War). After the expedition's dismal failure, a Scottish merchant and adventurer named Samuel Vetch campaigned for imperial officials to spearhead a major invasion of Louis XIV's Canadian strongholds.[49] Despite having largely ignored the Anglo-American war effort in the New World from 1702 to 1707, Whig-aligned authorities in London took Vetch's proposals seriously due to an increasing war of attrition on the battlefields of Europe and thanks to pressure from various interested merchant groups.[50]

By 1709, the British government approved Vetch's plans to drive the French from Montreal and Quebec and expected provincial naval forces to play a major role. Queen Anne herself ordered Vetch to ensure that forces from New York, Connecticut, and Pennsylvania build "six or more large Boats" and contract with Iroquois leaders to build canoes to help transport soldiers. Additionally, the queen requested that New England build various flat-bottomed transport vessels and that provincial authorities provide "able Pilots, whereof Captain Southweek [sic] to be one, & to go in his own Galley."[51]

That Queen Anne—or one of her royal officials acting in her name—knew Southack by name is unsurprising considering the queen's predecessor, King William, had personally rewarded Southack for effective privateering against the French in early 1693/4.[52] What is striking, however, is her outright support of and dependence on provincial and Indigenous naval resources throughout the northern colonies to support the proposed expedition against French Canada. Unfortunately for Vetch and his colonial partners, imperial authorities canceled the 1709 joint expedition without warning when peace talks seemed likely with Louis XIV that summer.[53] Vetch protested that preparations had been costly and that "our transports, flatt-bottom'd boats, whale-boats, as well as our troops being all ready att 12 hours warning; and because the fleet is so long a coming that the lateness of the Expedition may endanger some of the ships in their return to be blown off the coast."[54]

To recruit greater royal interest in a new expedition, Anglo-American and Iroquois diplomats traveled to London and convinced a new Tory administration in 1710 to recommit to Vetch's plan.[55] Royal officials sent two frigates and a bomb vessel from England to assist provincial forces in taking Port Royal, and three station frigates from New York and Boston joined the attack in early 1710.[56]

For the first time on the North American continent, significant provincial and Royal Navy forces assaulted a major target together. The fact that provincial naval forces would defer to Royal Navy command is evident when Dudley and his council advised that *Province Galley* be "Disposed in the fleet at the Direction of the Commadore so Soon as they Shall be ready to proceed."[57] Interestingly, Royal Navy commanders themselves asked for specific provincial naval support. For example, the captain of HMS *Dragon* asked the Massachusetts council for a local sloop to act as a "tender" to the Royal Navy vessels on the expedition.[58] All told, over thirty provincial transports with 3,500 troops from New Hampshire, Rhode Island, Connecticut, and Massachusetts joined Royal Navy vessels and *Province Galley* in the successful capture of Port Royal in the autumn of 1710.[59]

While provincial and imperial authorities alike rejoiced over the successful campaign, such luck was not guaranteed for every large joint expedition however. The following year, another Anglo-American campaign against Quebec failed after a major storm destroyed much of the royal and provincial fleet.[60] Even before the storm, the disastrous operation was rife with desertion and internecine disputes between royal and provincial military officials over supplies and perceived dedication to the expedition.[61] In particular, royal military officials accused New Englanders of not providing sufficient naval support for the expedition—a barb more often fired by provincial authorities at the Royal Navy than the reverse.[62]

While local vessels were vital in attacking enemy ports, they were also necessary to defend their own territories that came under siege by the French and Spanish themselves. One of the clearest cases of this phenomenon occurred in Charles Town, South Carolina, in the summer of 1706. A Franco-Spanish invasion force—emboldened by news that a yellow fever epidemic had weakened Charles Town's defenders—launched a major assault on the seaport.[63] It is uncertain what level of resistance the

Franco-Spanish invaders expected, but they were likely aware that the colony was a proprietary English colony without a Royal Navy guard ship.[64]

With an advanced warning from a New York privateer in the area, Governor Nathaniel Johnson organized a council of war, readied militia forces on land, and the "Vessels that lay in harbour were ordered to be fitted (viz) three ships one Briganteen & two Sloops + a fire ship." The fact that Johnson "ordered" these vessels to defend the colony implies they were impressed on the spot. Johnson commissioned Rhett as a vice admiral of this emergency fleet, and Rhett "hoisted the Union Flag on board the Crown Galley."[65] Rhett's usage of that flag is particularly noteworthy considering its legal and political implications. Under English law, only Royal Navy vessels could fly the Union Jack. Merchant vessels were limited to flying a similar banner with a "white escutcheon" in the center.[66] By flying the "Union Flag," Rhett flouted imperial law but perhaps intended to represent his makeshift fleet as the legitimate substitute for distant English forces.

Ultimately, Rhett's makeshift fleet of impressed merchant ships was successful. Upon seeing Rhett's hasty armada, the Franco-Spanish fleet retreated "in great hast + Confusion" without any resistance. Soon thereafter, Rhett took command of both the New York privateer ship and a local sloop to chase off scattered Spanish vessels. Even with the impressment of the aforementioned merchant ships, more volunteers joined Rhett in this final assault. One contemporary bragged of the "severall Gentlemen and others who were willing to share in the Danger and [honor] of that design." Rhett's naval forces were so successful that another observer boasted that with the "Providence of Almighty God," the colony's foes "like a Second Spanish Armada" met with destruction before the "flourishing colony."[67] While not every emergency fleet in colonial America stopped enemy invaders with such ease, Rhett's straightforward success against a joint European invasion with mere merchant vessels and militia forces attests to the utility of makeshift provincial naval defenses throughout the first two imperial wars.

Overall, the face of Anglo-American naval defenses changed dramatically throughout the more than twenty-four years of nearly continuous warfare between 1689 and 1713. Expanding upon their previous experience fitting out solitary coastal defense vessels during emergencies, colonial governments from New Hampshire to Barbados built semipermanent guard vessels and impressed emergency fleets to conduct diplomacy, hunt pirates, assist the Royal Navy, besiege enemy settlements, support infantry operations, and defend ports from imminent attack throughout King William's and Queen Anne's Wars. While these provincial navies were useful for military purposes, these expensive fleets would also amplify internal colonial controversies and challenge the fragile relationship between periphery and center in the nascent British Empire.

THE SOCIAL AND ECONOMIC COSTS OF PROVINCIAL NAVIES

Provincial navies often had social, political, and economic costs beyond what any colonial government had anticipated or expected. Although provincial governments built local fleets with the ostensible goal of defending their coastlines, the expenses and stresses associated with naval warfare amplified internal disputes over taxation, religion, race, and class. On a larger scale, provincial overreliance on local naval assets and the Royal Navy's inadequate protection of Britain's possessions in the New World highlighted larger weaknesses in the imperial-provincial military relationship.

Perhaps fewer historical examples better highlight the wider sociopolitical ramifications of provincial naval defense than the dramatic fate of Andros's provincial navy after the Glorious Revolution of 1689 and the even more eventful story of his provincial sloop *Mary*. In June 1689, not long after New England rebels deposed Andros and dispersed his military forces, pirates began to attack local commerce. Provisional governor Bradstreet and other officials ordered that "one Suitable Vessel be forthwith fitted out to Clear our Coast of Pyrats, which may be after Improved to transport Souldiers, Ammunition, and provisions for the Eastern Expedition, and from thence to Range the Coasts of Arcadia to Secure our fishing Vessels."[68] Because Boston authorities refused to restore the local Royal Navy guard ship (due to its captain's alleged Jacobitical and pro-Andros sentiments), they decided to use Andros's remaining provincial vessels to hunt pirates.[69]

In August, provincial officials decided to send Captain Joseph Thaxter with the provincial sloop *Resolution* to go hunting for a pirate named Thomas Pound who had captured two vessels off the coast.[70] Intriguingly, Pound was the former captain of *Mary* and had even served as a pilot for the now-deposed Royal Navy captain George. It is worth noting here that provincial authorities rotated captains and officers of provincial ships on a frequent basis. After an initially unsuccessful hunt, in late September provincial authorities sent Captain Samuel Pease and the crew of *Mary* to search for the same vessel's former-captain-turned-pirate.[71] Pound's forces, confident of their prowess, sent word to Boston via one of their victims that they would slaughter the entire crew of any "government sloop" sent out against them.[72]

By early October, Pease and *Mary's crew* discovered and overpowered Pound's pirates in a vicious melee near Martha's Vineyard. During the fierce battle, Pease was mortally wounded. While provincial authorities could have executed the entire pirate crew after this bloody engagement, the Boston court only executed the pirate responsible for killing Pease and spared Pound and the rest of his men.[73] New England divine Increase Mather praised that "small Vessel of Brisk *Bostoneers*, who in Their Majesties Name and under Their Colours, maintained a Bloody Fight with the Rogues and took them" but also alleged that George of HMS *Rose* supplied the pirates with ammunition. It is not clear if this accusation was true, but some contemporaries including one of the accused pirates substantiated this claim.[74]

At the start of the new decade, New England agents defended their political revolution against Andros before the new king in London, William III, and also traded barbs with their former provincial overlord over his handling of the region's coastal defense. In an undated letter from Bradstreet to the king, the aged governor detailed his plans to send Andros back to England, asked for the restoration of Massachusetts's original charter, and briefly noted that he had been "necessitated to grant Commissions to suppress, bring in and secure" Pound and other pirates.[75]

In contrast to Bradstreet's promise that he was doing all that he could to hunt pirates, Andros blamed the new Massachusetts government for coastal insecurity. He took responsibility for the initial creation of the colony's provincial navy, including *Mary* and *Speedwell*, along with two private vessels before the "subversion" of his regime. Andros complained that the rebels dispersed his forces, which led to Franco-Indigenous incursions that endangered the lives of Anglo-American colonists on the northern borderlands, the fisheries, and even the New England forests that helped supply raw materials for the Royal Navy. Without King William's intervention, Andros argued that Franco-Indigenous forces would overpower those northeastern colonies, which lacked "Provisions ... Ships, Vessels, Seamen, and other Necessarys ... to supply or Transport any force."[76]

Massachusetts's agents in London contested Andros's criticisms and insisted that Andros himself had mismanaged provincial naval forces during his controversial reign. They alleged that one of Andros's Catholic military officers "had [been] suspected to be in a Plott for deserting and runing [sic] over with the Sloop *Mary* to the French." They further accused Andros of having impressed private vessels for inane tasks and claimed that he had not paid his sailors. These accusations highlighted

Massachusettsans' growing anxiety over a suspected Franco-Catholic conspiracy to destroy their godly commonwealth.[77]

While Massachusetts's agents resented what they believed were Andros's lies, they would soon face a much more material challenge connected to the revolt of 1689. New York's newly appointed governor, Henry Sloughter, insisted that one of the Boston government's two publicly funded sloops—likely referring to *Mary* and *Speedwell*—should be given to his colony since Andros had commissioned them under the guise of the Dominion of New England—the megacolony that New York had just recently been a part of. Based on the advice of the Lords of the Committee for Trade and Foreign Plantations—the future Board of Trade that would handle colonial affairs—in April 1690 King William ordered that one of the publicly funded sloops be sent to Sloughter.[78]

What ensued was a lengthy transatlantic argument over who owned the first provincial navy that Andros employed in 1688, whether they were publicly funded by taxpayers or not, and which vessels were in service by the time King William intervened in these disputes in early 1690. The sources are often too contradictory or not clear enough to make sense of.[79] Perhaps this is unsurprising considering the sociopolitical chaos that has followed political revolutions throughout history. Nevertheless, by 1691, Governor Sloughter's successor in New York had planned to send one of the Royal Navy's hired sloops, *Archangel*, to go and seize the sole remaining sloop from Andros's fleet—*Mary*.[80] While it appears that this seizure never happened, it is worth noting that extended debates over small provincial warships not only complicated the fallout from Massachusetts's experience during the Glorious Revolution but nearly led to bloodshed between New Englanders and naval forces from New York.

If Massachusetts authorities thought their legal troubles with the sloop *Mary* were over, they were sadly mistaken. While *Mary*'s former captain Pound had turned to piracy, her new captain, John Alden, faced charges of witchcraft in the Salem witchcraft trials of 1692. In the years leading up to the witchcraft crisis, provincial authorities entrusted Alden with various missions on the northern borderlands, including helping to free captives from Franco-Indigenous forces. Critics claimed that Alden had done little to help captives and only wished to trade with New England's enemies. This became especially apparent when Alden fled with ransom money for captives held by French authorities in 1691 and when he attacked a French vessel despite being granted safe passage by French negotiators. Alden's selfishness led to the continued captivity of various provincial officials, including his own son John Alden Jr. It is possible that this corrupt behavior, coupled with accusations that Alden was responsible for an Indigenous attack on York, Maine, (and other miscellaneous charges including the fact that he had Indigenous lovers) inspired girls in Salem to accuse him of having a leading role in satanic rituals alongside accused warlock and former minister George Burroughs.[81]

Though misdeeds on the northern borderlands may have inspired some of the accusations against Alden, historians have long seen the accusations against Alden within the context of larger provincial fears of incompetence or malevolence among the region's leaders during the turbulent early years of King William's War. Paul Boyer and Stephen Nissenbaum make the case that accusations against Alden, who was one of the "best-known men in New England," came at a time when the girls in Salem were starting to accuse sundry provincial leaders of sorcery.[82] Mary Beth Norton argues that many of the accusers at Salem were childhood survivors of massacres in King Philip's War and "saw Alden's collusion with the Wabanakis, devil worshippers who had devastated their families, as an indication of his fidelity to Satan."[83] Louise A. Breen contends that Alden's misdeeds while entrusted with his very real position of military authority coupled with his alleged role as an officer in a spectral legion of evil spoke to a growing "elite fear of pacts with Satan that could endanger the civil state" of New England.[84] All told, Alden's abuse of his position of authority within the colony's provincial navy led to fears that the region's coastal and spiritual security were both compromised. This fear was evident even before the 1692 witch craze, when the provincial government decided to impose restrictions on his voyages with *Mary*, including forbidding him from transporting trade goods or extra ammunition that might be given to the enemy.[85]

During the trial, Alden put up a fiery resistance to both the judges and his accusers—some of whom were young girls and women descended from victims of Indigenous raids. When the accusers claimed that he made them fall to the ground by looking at them, Alden boldly asked the judge, Bartholomew Gedney, why he did not face a similar fate when he looked at him. Despite his initial plan to resist the witchcraft charges in the court room, Alden made the wise choice of escaping from confinement, weathered out the trials, and lived to be eighty years old.[86]

Despite Alden's ultimately happy fate, mistakes made during his provincial naval service contributed to the dramatic and lethal climate surrounding the infamous trials. Ultimately, from 1689 to 1692, Massachusetts's small provincial navy played an oversized role in amplifying colonial disputes surrounding the Glorious Revolution, battles with pirates, intercolonial rivalries, and even the Salem witchcraft trials. These examples showed the widespread impact provincial navies played in the first major imperial contests of this era.

If the sociopolitical ramifications of provincial naval warfare during the imperial wars were high so were the economic costs. In fact, the upkeep of a provincial warship could be just as expensive as the costs of paying a provincial militia unit. For instance, on October 3, 1704, Governor Dudley and his council ordered the colony's treasurer to pay £191 to Captain Nathaniel Jarvis and the crew of the brigantine *John & Abiel*—a private vessel that had been hired as a "Vessel of War, in the late Expedition into the Bay of Fundey." This accounted for 113 days of crew wages and vessel hire costs between April and August 1703. The sum was reduced to £167 to account for supplies taken from the commissary. That same day, Dudley's council ordered the treasurer to pay a militia company ninety pounds for a month of wages between June and July 1703.[87]

While it is true that the government paid the infantry company more money for a shorter service period, one must also consider that the colony had to pay for the constant upkeep and repair of its provincial warships as well. For instance, in September 1701, Dudley and his council ordered the treasurer to pay Southack and various Boston businessmen £294 for "materials as cables, Sails, a new Boat . . . for his [Majesty's] Ship the *Province Gally* . . . and for workmanship of Carpenters and others in fitting said Ship . . . and Provisions for victualling the same."[88] Ultimately, the construction, hiring, upkeep, and provisioning of provincial naval vessels could rival if not exceed the costs of maintaining provincial regiments on land.

The Barbadian government's troubled attempt to keep a flotilla of guard ships in 1702 and 1703 highlighted the economic and social woes a large provincial navy could bring even in the wealthiest of England's sugar colonies. In August 1702, the Barbadian assembly resolved "that a levy of 6d. a head on negroes, be raised for a fund for setting out ships of war, and also that 6d. per tun on every ship arriving to this island shall be levied." The next day the Barbados Council and Assembly together agreed to pass an act to encourage privateers and to impress guns and men "for fitting out two vessels of war" and believed it was "lawful and justifiable, it being" for Her Majesty, Queen Anne's service.[89] Provincial officials proceeded to compile an impressive fleet of four galleys, sloops, and brigantines.[90] By early September, even as both legislative houses of the Barbadian government contemplated impressing men from privateer vessels because of manpower shortages, the assembly paid for another vessel—the brigantine *Larke*. Around that same time, provincial officials fired Captain John Smith from his role as skipper of the provincial sloop *Constant Jane*. His sailors complained that he had beaten them during an attempt to impress them into provincial service.[91]

With rising costs, dismissals of officers, and complaints of intraservice abuse, it was becoming apparent that this provincial navy brought more woes for the Barbadian government than it was worth. As if insufficient manpower and the abuse of sailors were not problematic enough, a week later the Barbadian Council and Assembly learned that *Constant Jane* had shipwrecked. Some in the government came to suspect that she was "wilfully run on shore by Thomas Driffield, Lt. of the vessel, and others" and initiated an investigation.[92] To add insult to injury for the provincial government, by the end of September the crew of the brigantine *Madeira* mutinied and ran away with the ship.[93]

Nevertheless, despite these setbacks, the provincial government pushed on with matters of defense. Various assemblymen volunteered personal funds to repair a provincial vessel, and one official volunteered his own sloop to carry a warning about French privateers to a Royal Navy ship cruising with one of the island's brigantines. Civic volunteerism could be costly however, and the assembly filed a petition to London to consider the "growing charge of fitting out vessels of war to [Her Majesty's] service."[94]

Despite the fact that provincial navies amplified sociopolitical and economic crises within the British colonies, the Royal Navy's meek presence in the New World continued to force colonial governments to rely on provincial naval defense. Throughout King William's War and much of Queen Anne's War, royal military assistance to the colonies (particularly outside of the West Indies) had been extremely limited. Nevertheless, as seen above, even those few Royal Navy guard ships in colonial seaports during this era did not guarantee coastal security. Disputes between provincial authorities and Royal Navy captains could break out over several issues, including traditional battles over the chain of command, the business ventures of Royal Navy officers outside the parameters of their military duties, and the ever-controversial issue of impressment.[95]

Arguments over Royal Navy impressment policies in particular would remain a major cause of provincial–Royal Navy tensions for decades even after Parliament's passage of the America Act of 1708 (a.k.a. the Sixth of Anne). With pressure from Caribbean merchant captains who lost untold numbers of sailors to Royal Navy press gangs, Parliament decided to act and limited Royal Navy impressment lest it damage lucrative Caribbean commerce. The Royal Navy was forbidden from impressing merchant sailors and privateersmen in the New World.

While the legislation may have been intended to ease provincial tensions with Royal Navy commanders, the Sixth of Anne created more problems than solutions for Royal Navy manpower. On the one hand, the law essentially limited Royal Navy commanders to the initial crews they left England with. On the other hand, the law did not specify whether provincial governors had the right to impress men for provincial naval service or for Royal Navy ships when requested. In response to numerous provincial queries over whether colonial governors had the right to initiate impressment, the Board of Trade gave inconsistent and vague answers. Westminster's silence over the full extent of the ban encouraged the Admiralty to ignore the prohibition and to continue allowing its officers to impress at will by the 1720s.[96] While the Royal Navy's leadership took until the 1720s to reinstate its impressment policies, provincial governments had never truly stopped the impressment of men and vessels into colonial service.

Aside from disagreements over impressment, personality conflicts between provincial governors and Royal Navy officers often soured an already poor working relationship between colonial and royal military officials. Once again, a dramatic encounter involving the provincial sloop *Mary* serves as an illustrative example of growing tensions between provincial and Royal military leaders in this era. When King William appointed Phips—the veteran general of the 1690 Quebec expedition—as Massachusetts's new governor in 1692, he gave him two Royal Navy station ships—HMS *Conception Prize*, captained by Robert Fairfax, and HMS *Nonsuch*, captained by Richard Short. The captains and the governor disputed over joint failed business ventures, locations for coastal patrols, and the provincial government's material support for the royal ships. The breaking point in this strained relationship came when Phips asked Short to send Royal Navy sailors to serve on the provincial sloop *Mary*, and Short refused to help crew the provincial vessel. A physical altercation broke out between the two men on January 4, 1692/3 that would ultimately lead to Short's imprisonment by provincial authorities and the royal government's eventual dismissal of Phips from his office.[97]

While disputes between provincial and Royal Navy officers damaged the military partnership between periphery and center on a microlevel, inconsistent messages from London regarding future royal military intervention would foster confusion on a macrolevel. Although the Crown did slowly increase royal military intervention in some sectors by the end of Queen Anne's War, it never abandoned its "insistence on colonial military self sufficiency."[98] This ethos, along with the still-limited nature of royal military assistance encouraged Anglo-Americans to continue to rely on their own provincial navies.

Ultimately, throughout King William's War and Queen Anne's War, Anglo-Americans came to depend more and more on temporary and semi-permanent provincial navies to secure their coasts and to wage offensive campaigns against enemy ports. While these forces were useful for immediate defense needs, their social and economic costs could outweigh their military utility at times. Despite these setbacks, the Royal Navy's still-weak presence in the New World forced Anglo-American governments to continuously rely on provincial navies to preserve their coastal trade and the safety of their ports. These forces—and the increasingly uneasy relationship between royal and provincial military officials—would be put to the test again in the irregular maritime conflicts that followed the end of Queen Anne's War in 1713.

3

Provincial Navies and Irregular Warfare on Imperial Borderlands, c. 1713–1739

When Great Britain and its foes signed the 1713 Treaty of Utrecht that ended the War of the Spanish Succession, there were plenty of reasons why imperial officials hoped to avoid future wars with the French and Spanish. With the British South Sea Company's newly acquired rights to trade enslaved Africans to Spanish colonies coupled with the profitability of illicit trade with Spanish colonists, imperial planners in London discouraged aggression against the Spanish in the Americas to protect these fragile new trade avenues. Additionally, even though Queen Anne's government had expanded its financial borrowing and taxation powers during the war, Great Britain's coffers were drained by the enormous costs of the conflict.[1]

Despite Britain's hope for respite after years of costly combat, a perfect storm of foes emerged to challenge any pretension for peace. Spanish colonial authorities—ever concerned with British interloping in their costal trade—hired *guarda costas* to seize British and Anglo-American vessels with suspected illicit trade goods on board. Far from merely enforcing trade laws within Spanish territories, *guarda costas* frequently employed noteworthy violence in their seizures of merchant vessels. While some of these *guarda costas* had legitimate commissions from Spanish governors, Anglo-Americans suspected that many of the captains feigned official support in order to justify outright piracy. The most famous outrage over *guarda costa* violence came in the early 1730s, when Spanish mariners cut off the left ear of suspected smuggler Robert Jenkins. This episode would serve as a major cause of the next lengthy conflict between Spain and England, the aptly named War of Jenkins' Ear.[2]

Even though Britain and Spain were able to avoid an all-out conflict for the two and a half decade period stretching from 1713 to 1739, there were short episodes of open warfare between the two empires from 1718 to 1721 and from 1726/7 to 1729. While the threat of an expanded international conflict loomed large during these short wars, British forces never undertook serious campaigning in the Western Hemisphere, and most of the brief fighting was limited to European battlefields. Ironically, it would be during a time of tacit peace in the early 1730s when Britain made its most successful move against Spain in the New World: the creation of the Georgia colony.[3]

As if the ever-present specter of the return of imperial war was not troublesome enough, the years following Queen Anne's War also saw a major wave of Anglo-American piracy. Without consistent military employment after 1713, thousands of British and Anglo-American sailors turned to illegal maritime raiding against their former Spanish enemies. With rising opposition to illegal swashbuckling in formerly welcoming colonial ports, Anglo-American pirates also began to attack their own countrymen to fund their "trade."[4] From 1715 to 1725, during what historians have come to call the Golden Age of Piracy, thousands of these renegade British sailors and mariners from other nations would use the weakly governed Bahamas as a base from which to pillage and plunder throughout the Atlantic world.[5]

One final problem also challenged Britain's imperial aims in North America after 1713: major terrestrial and maritime threats from powerful Native American nations on colonial borderlands. For nearly a decade after their French allies ceded some of their Acadian lands to the English, Wabanaki mariners waged their own naval war against Anglo-American colonists well into the late 1720s.[6] Around the same time, Anglo-American officials in South Carolina faced terrestrial and maritime attacks by aggrieved Yamasee fighters after years of Carolinian trade corruption and enslavement of their Indigenous neighbors.[7] The Yamasee onslaught was so fierce that it came close to nearly destroying the colony of South Carolina itself.

This tense era also highlighted the continued pitfalls of Royal Navy intervention in the Americas. To be certain, the Royal Navy did make a number of important advances following the Treaty of Utrecht. Developments during this period included various financial innovations, an expanded number of expert administrators within the First Lords of the Admiralty, the growth of naval yards throughout the empire—including at Jamaica and Antigua—and more efficient methods of supply procurement.[8] Additionally, in the early 1720s, the Royal Navy expanded its fleet of agile sloops in the West Indies to counter threats from pirates and *guarda costas*.[9]

Notwithstanding its many administrative advances, internal political controversies within Britain and the wide array of irregular threats throughout the Atlantic world limited the Royal Navy's effectiveness in the interwar period. At home, opposition to the Walpole administration grew after his lackluster and nonaggressive utilization of the Royal Navy during a short war with Spain in the late 1720s.[10] Outside of executive indecisiveness, the Royal Navy could not possibly combat the British Empire's vast array of enemies all at once. Historian Eliga Gould makes that case that while the British Empire was effective in curbing maritime piracy in the 1720s, its fights with *guarda costas* and Indigenous forces on the continent "underscored the limits on Britain's ability to enforce its agreements with other European governments, one along the inland reaches of North America, the other in the coastal waters and shipping lanes of the Caribbean and the Western Atlantic."[11]

While the contingencies of King William's and Queen Anne's Wars forced Anglo-American governments to create large-scale provincial navies, Royal Navy inattention and postwar troubles on the British colonies' ill-defined edges forced the governments on imperial borderlands (namely Nova Scotia, Massachusetts, South Carolina, and various West Indian governments) to continue those naval establishments and to develop flexible—and largely independent—naval responses to irregular threats from Indigenous, piratical, and Spanish foes.[12]

Provincial Navies and Imperial Borderlands: New England and Nova Scotia, 1715–1728

In the wake of Queen Anne's War, violent clashes with Native Americans on the South Carolina and Acadian/New England borderlands forced Anglo-American officials on opposite ends of the mainland colonies to utilize provincial naval forces in similar ways. These border wars coincided with the ongoing golden age of Anglo-American piracy, and it was not uncommon for officials in both regions to have to navigate a complex of Indigenous, piratical, and traditional imperial threats all at the same time.

While Anglo-American officials in both regions continued to prefer elusive royal military assistance, imperial authorities did little to ensure adequate Royal Navy protection for its many North American ports in the years following Queen Anne's War.[13] New England's (and by extension Nova Scotia's) maritime expeditions against Indigenous and piratical foes during this period demonstrated growing provincial naval self-reliance in the wake of inadequate royal protection. This is not to say that provincial officials eschewed royal assistance en total. For instance, when a pirate vessel was spotted off the coast during the spring of 1717, Massachusetts's governor dispatched "Capt. Cayley of His Majesty's Ship *Rose*, and Capt. Coffin in a Sloop well Arm'd and Man'd with 90 Men to go out in quest of the said Pirate."[14] Not long thereafter, however, the colony's House of Representatives voted to continue the sloop "in the Service for the Defence of the Coast" *until* the next Royal Navy ship was to arrive.[15] The Massachusetts legislature was willing to pay for a provincial sloop but hoped to delegate the responsibility of naval defense on a Royal Navy frigate if possible.

Notwithstanding the preference for Royal Navy assistance, Anglo-Americans in Nova Scotia and New England found themselves largely alone in their borderland conflicts with the Wabanakis. After the British capture of Port Royal (later Annapolis Royal), Nova Scotia, in 1710, imperial officials had to contend with the administration of hostile French Acadian colonists and various Wabanaki tribes including the Mi'kmaq, Abenaki, and Maliseets. This dilemma occurred while the imperial government and provincial authorities were still dealing with military threats from French authorities.[16] When French officials deeded much of their former Acadian colony to the English at Utrecht in 1713, angry Wabanaki leaders initiated a decade-long maritime war against their Anglo-American neighbors throughout the coastline stretching from Newfoundland to Maine. Historian Matthew Bahar contends that while the Wabanaki raided English vessels in order to preserve their regional hegemony, English officials considered Indigenous assaults on English shipping in the region to be outright "piracy."[17]

By the early eighteenth century, the Wabanaki confederacy had nearly two centuries of experience in operating European vessels. As early as the sixteenth century, Wabanaki fishermen and mariners captured small sailboats called shallops that had been abandoned by European explorers. Throughout the following centuries, Wabanaki mariners stole or purchased similar small craft and employed them in raiding or in trade missions. Interestingly, by the mid-seventeenth century, some Wabanaki naval officers even started to don European gentlemen's clothing to assert their social status as leaders of naval crews.[18]

By appearance, these raids mimicked the ongoing pirate scourge in the Caribbean. After all, the Wabanaki mariners used light craft ranging from canoes to better armed shallops and sloops to swiftly move on their English prey.[19] While shallops and other small craft were the preferred craft of Native naval forces, their colonial pursuers in New England's provincial naval forces often had the same sort of craft. A recent study's description of Massachusetts's "hulking, heavily armed, and consequently slow warships ... [that] failed in their pursuit of more agile Indian mariners" overemphasizes the differences between both sides' vessels.[20] For instance, in 1723, a militia leader named Captain Heath led several men in whaleboats to ambush Wabanaki mariners in canoes. While many of the Natives escaped, the militiamen captured one "Canoo, one Gun, their Ammunition, and other stuff: the Canoo was shot through where the Indians sat."[21] At the end of the day, both Wabanaki mariners and their provincial naval opponents relied on the same sorts of small sail-and-oared vessels to pursue their prey.

While independent Wabanaki raids could be devastating, tacit Franco-Acadian support of these raids amplified Anglo-American anxieties for their coastal security. Thanks to territorial vagueness in the 1713 Treaty of Utrecht, both Anglo-American and French officials claimed to own the island fisheries between Nova Scotia and Cape Breton Island. In the treaty, the French had agreed to surrender Acadia—a region that they believed only included mainland Nova Scotia. Interwar disagreements over the status of the Canso fishery on the northeastern edge of Nova Scotia led to a state of near war between French and Anglo-American authorities.

After Massachusetts dispatched a Royal Navy frigate to destroy the controversial French fishery at Canso in 1716, Governor Saint-Ovide of Île Royale on Cape Breton Island—with support from the French Crown—encouraged Mi'kmaq forces to attack New England vessels in return.[22] One dramatic episode in 1720 highlighted the dual threat posed by Franco-Indigenous raiders on the northern borderlands. A report from Canso in the late summer detailed an attack by a "Company of Indians with some French assisting them." The raiders surprised the Anglo-American residents in their beds, stole their valuables, and transported the goods on French vessels. Even though the French governor at Cape Breton promised to prosecute any of his countrymen involved, the Anglo-American correspondent believed that there was a "plain Confederacy between the French and Indians, to ruin the people and fishery here." Subsequent interviews with French prisoners revealed that many of the Franco-Acadian sailors involved in the raid were fishermen angry over the loss of the French fishery at Canso.[23] By 1722, violent borderland tensions such as these would help fuel a four-year conflict with French-aligned Natives in the area known as Father Rale's War (so named for a renegade French priest) or Dummer's War (so called for the governor of Massachusetts after 1723).

During the conflict, provincial naval forces would be vital to securing Britain's feeble hold on its northernmost American coastal borderlands. In August 1720, Nova Scotia governor Richard Philipps was confident that he could save the British government significant money by hiring a sloop to guard the coast against Anglo-American smugglers attempting to covertly trade with the French.[24] Soon thereafter, he forwarded a petition from various colonists that described a merchant being forced to fit "out two small vessells in pursuit" of Franco-Indigenous robbers. The colonists had begged for "men, arms and ammunition to enable them to defend the 'rights of the Crowne of England'" and claimed that Native captains confessed to acting on official orders from the French-Canadian governor Doucet. Alongside this account of provincial naval struggles with enemy raiders, Philipps sent a standard plea for Royal Navy assistance.[25]

In this situation, Philipps both asserted that Anglo-Americans *could* defend their shores independently but also reiterated their desire for royal military aid. As ever, the royal government recognized the province's ability to handle naval defense and did its best to avoid excessive naval spending in the region. In December 1720, the Board of Trade suggested that Phillipps should "be allow'd according to his own proposal to hire a sloop for the defence of that coast and the preventing of illegal trade there."[26] Despite grumbling over delivery times from Boston shipwrights, Philipps did note that the "obtaining thereof [was] cheifly oweing to your Lordshipps."[27]

Even if the Board of Trade was influential in Philipps's obtaining a military vessel (the schooner *William Augustus*), imperial authorities were generally loathe to finance colonial naval defense efforts. Much of this is evident in Phillips's petition to the Board of Trade begging for financial compensation for the guard vessel's operating costs. According to Philipps, the Board of Trade asked Boston's Royal Navy post captain Thomas Durell to survey coasts around Nova Scotia and Placentia. Durell said that such a thing was "impracticable with [His] Majesty's ship under his command and advised that a small vessel might be built at Boston. This the Governor [Philipps] was instructed to do, and gave a letter of credit to Capt. Durell, who contracted for it at Boston."[28]

Even though the Lords of the Treasury were of the "opinion that the Governor Col. Philips's charges should be reimbursed by the Navy," the Lords of the Admiralty argued in 1724 that the navy was not responsible for the sloop's costs and declared that they would not assist Nova Scotia with this project because "when vessels have been fitted out by the Governors of his Majesty's Islands or Plantations abroad, the inhabitants have [always] borne the charge thereof." They did, however, make the meek promise that when Royal Navy vessels were sent to guard the Newfoundland fishery that one of them would be "appointed to attend on Placentia and Nova Scotia" in winter months when it was too icy to operate in Newfoundland.[29] With that 1724 policy statement, the Lords of the Admiralty made it known that Anglo-American authorities were *allowed* to fit out their own provincial navies but that they were to fund and operate them with their own resources. This evasive statement would reflect the royal policy toward provincial navies for the next two decades.[30]

Nova Scotia's southern neighbors in New England also harnessed their own provincial naval power to simultaneously fight Wabanaki mariners and the ongoing scourge of Anglo-American pirates. As we saw in the previous chapter, these efforts were not always successful or free from contemporary critique. For example, in June 1722, James Franklin, a newspaper printer and older brother of future Founding Father (and naval advocate) Benjamin Franklin, was arrested by Massachusetts authorities for mocking the colony's many delays in fitting out a vessel to hunt the pirate Edward Low.[31]

These critiques have been echoed by modern scholars as well. For instance, Matthew Bahar has recently argued that it was only exceptional colonists who gathered enough fortitude and firepower to hunt Indians at sea" and that they quickly "became the hunted."[32] While colonial forces did make numerous tactical blunders in Dummer's War of 1722–26, Anglo-American authorities throughout the New England colonies did succeed in mustering several makeshift fleets to fight both Wabanaki and Anglo-American piratical enemies with limited Royal Navy assistance. While these naval expeditions had mixed results, the war against the Wabanaki was strained more by strategic limitations than any lack of "fortitude."

On June 6, 1722, Massachusetts governor Samuel Shute and his council discussed news from Rhode Island about a "Pyrate Vessel on the Coast" that had captured a vessel from Charlestown, Massachusetts. They ordered Royal Navy captain Durrell to take HMS *Seahorse* on a cruise to hunt for the "said Pyrate Vessel, and to guard and to Protect this Coast."[33] By that point, Rhode Island's government had already sent two provincial sloops after the pirate.[34]

Over the next few days, a joint committee from both of Massachusetts's legislative houses decided to expand the hunt against the pirate and voted to impress a sloop and appointed Captain Peter Papillion to lead the expedition. Aside from guaranteeing a month's worth of provisions and funding for one hundred men, the committee promised a fair share of the "Goods, Wares & Merchandizes . . . that Shall be found on Board. . . . So far as is Consistent with the Acts of, Parliament . . . And for Further Encouragement; That they be paid out of the publick Treasury" ten pounds for every pirate killed or captured, as well as insurance for possible injuries.[35] The colony's ability to raise a sloop and raise a crew with such speed while also funding expeditions against Natives was only possible because of its continuous circulation of paper money.[36]

This growth of the web of maritime operations is evident in a June 20 letter from Archibald Cumings, a Boston custom officer, to William Popple, the secretary of the Board of Trade. Cumings reported that the "government of Rhode Island, fitted out two Sloops" to pursue two pirate vessels, and Massachusetts sent a provincial ship's crew to join them because the Royal Navy station frigate was away patrolling in the Canso fisheries. In a postscript, Cumings remarked that Massachusetts had also deployed "200 Men at the Eastward and are Sending 100 more as an Additional force" to fight Wabanaki mariners.[37] Cumings's letter hints at the multifaceted New England naval war against both Anglo-American pirates and Wabanaki sailors.

While these efforts kept the colony busy, Massachusetts's large net of naval operations still only grew wider from here. A week after Cumings sent his letter to the Board of Trade, another legislative committee met to discuss the specifics of the campaign. Among the naval recommendations of the committee were that "Ten Whale Boats with very good Oars be provided, and sent to the Forces, for Enabling them to manage a sufficient Scout" and that a "Sloop be taken into the Province Pay for Transporting Men and Provisions." Over the next few days, the governor and assembly extended Captain Papillion's pirate hunt for a month and ordered Captain Durell to patrol as well with HMS *Seahorse*.[38]

By late July, Massachusetts authorities also decided to dispatch two sloops as far north as Nova Scotia to search for Wabanaki mariners who had kidnapped New England fishermen.[39] The following month, Durell, who had previously suggested Nova Scotians utilize provincial forces to defend their coasts, offered to man small provincial vessels with Royal Navy sailors to hunt down the same foes.[40] He likely realized that his own warship was too large to pursue the swift Wabanaki light craft and tapped into provincial naval resources to supplement his own mission to defend the coast.

It is noteworthy that during this conflict, we find one of the rare recorded occasions where Anglo-American women played a role in organizing naval expeditions. On June 27, 1722, two civilians named Christian

Newton and Margarett Blin (also spelled Blyn) informed the House of Representatives that they had fitted out a sloop and crew to recapture loved ones taken by Wabanaki forces and requested arms from the provincial government for their crew. The next day, a committee from both houses agreed that thirty soldiers "under a proper Officer (whom His Excellency [Samuel Shute] be desired to Commissionate) with Provisions, Arms and Ammunition to be put on Board the Sloop offered by *Margaret Blin* . . . to repair as soon as may be to *Passmaquada*, and there to use their best Endeavours to recover [captives] from the Indians." They also suggested that the militiamen capture Indigenous captives if they could not liberate the New England captives.[41] Ultimately, this rescue mission never commenced as Margarett's husband—the sloop captain James Blin—and the other captives made a successful escape.[42]

Despite the vast array of Massachusetts forces sent out after them, Wabanaki sailors would capture scores of English vessels and kidnap large numbers of Anglo-American colonists throughout the summer. This was only the beginning of what would be a four-year onslaught that would see Indigenous chiefs leading formidable fleets—including flotillas of a half dozen sloops and schooners—against New England and Nova Scotia mariners. In some cases, Wabanaki seafarers found a willing market for Anglo-American vessels at the French fortress of Louisbourg. For French authorities, Wabanaki raids on English shipping damaged their imperial competitors without requiring overt French involvement.[43]

While provincial naval forces and detached militia "marines" onboard local vessels were able to score some important victories by the end of 1722, including securing the Canso fishery, the Wabanaki confederacy continued a devastating offensive by land and sea on the northeastern borderlands for years to come. Even after New England forces assassinated the Wabanaki's major Acadian ally, Father Sebastién Rale, in 1724, Maliseet fighters still leveled many Anglo-American homesteads on the Maine frontier. By 1725, however, Nova Scotia and Massachusetts emissaries made diplomatic headway when they threatened French officials at Montreal and Louisbourg with a general assault on Franco-Acadian shipping and settlements if they did not cease their support of the Wabanaki war effort. The threat of a new conventional European conflict in the region coupled with growing dissension within the Wabanaki ranks forced the French officials' hand on the matter. In 1726, war-weary Anglo-American officials and their Indigenous enemies agreed to separate ceasefires in Nova Scotia and New England.

All in all, Father Rale's War was over, yet without either side having a true strategic "victory."[44] Any successes made by Anglo-American forces during the conflict depended on various New England authorities' consistent deployment of transport sloops, guard vessels, whaleboats, along with occasional cooperation with the overworked Royal Navy captain Durell to pursue Anglo-American pirates—a threat that had never truly dissipated, even as the Massachusetts government directed most of its military attention to the fight against the Wabanaki.[45]

Unfortunately, while Durell had been a dependable partner for regional leaders, his eventual successor, Captain James Cornewall of HMS *Sheerness*, was less cooperative with provincial forces. In late June 1726, Lieutenant Governor Dummer and his council refused to allow Captain St. Lo of the undermanned frigate HMS *Ludlow Castle* to impress New England mariners. Not long thereafter, the council received reports about the Anglo-American pirate William Fly raiding nearby shipping. In response, Dummer's council ordered that "a good Sailing Sloop or other Suitable Vessel be taken up for his [Majesty's Service against the Said] Pirate." The council appointed William Atkinson (himself a victim of the pirate) as the captain and established a pay table of eight pounds a month for the captain, payments for his officers in "proportion," four pounds for able sailors, a twenty shilling bounty for volunteers, and the usual promises of insurance for the wounded.[46] The governor would later boast about the "cheerful and ready appearance of [forty] volunteers upon the bounty offer'd for that service" within six hours of the commencement of the recruitment drive.[47] The sloop *Loyal Heart* was ready to sail.

The next day, Dummer and the council dismissed Atkinson due to suspicions he had associated with the pirates and replaced him with Captain Thomas Little. They also discussed "threatening" letters that the acting governor Dummer had gotten from Captain St. Lo regarding the attempts to outfit a pirate-hunting sloop. They were shocked to find that fellow royal captain Cornewall had stopped *Loyal Heart* with HMS *Sheerness* in the middle of Boston Harbor.[48] According to a subsequent complaint by the Massachusetts governor to the king, Cornewall had demanded to know their business and threatened to fire on them. When Captain Little told Cornewall that he had a provincial commission to hunt pirates and tried to continue his voyage, Cornewall ordered his sailors to fire on the little provincial sloop. Dummer and his council complained that this behavior was "very far from answering your Majesties gracious intentions" in providing Royal Navy protection and complained that Cornewall had done little to actively defend the coast for the last two years. They asked for a new station captain and requested that governors have more control over future Royal Navy officers.[49]

Of course, Cornewall's narrative of the events was notably different. In *Sheerness's* logbook entry for July 1, 1726, Cornwall recorded that that a "Sloop hauld of [sic] from the Wharfe . . . Armed with 6 Guns & 4 Patterreroes, & as near as I could Guess about 40 hands, So unexpected a Sight could not but be Very Surprizing to me having not the Least Infirmation." Cornewall claimed that the sloop's master ignored calls to answer a summons to board *Sheerness* and proceeded to sail away. This prompted Cornewall, who believed them to be "going a Pyrating," to fire four times on the sloop, which anchored near the safety of Castle William with only limited damage to its sails.

The drama was far from over however. On July 22, Cornewall noted that the "Sloop Said to be fitt'd out at the Expence of this [Government] Arriv'd here & this Morning . . . hoisted a King Jack." As will be recalled, this flag was solely reserved for Royal Navy vessels, but provincial warships frequently flouted this rule during operations. When Cornewall sent his men to forcibly take the King's Jack down, a brawl occurred, and thirty provincial sailors with pistols and swords forced the Royal Navy men back.[50]

While Cornewall's behavior did not permanently sour provincial opinions of Royal Navy assistance, it was certainly part of a growing pattern of what historian Douglas Edward Leach has described as a "kind of self-assured arrogance that naval authority seemed to generate and that was so offensive to many colonists."[51] Colonial resentment was certainly evident in late August 1726, when the governor's council and the colony's legislature both commended *Loyal Heart's* Captain Little for "*having handsomely Asserted and Defended the Honour of this His Majesty's Government of this Province, and of the Commission he had born . . . notwithstanding the Violent Opposition given him by Capt. James Cornwall.*"[52]

In the minds of New England and Nova Scotia officials, their provincial navies had secured their coastlines from Indigenous and piratical threats for the "Honour of this His Majesty's Government" even when the Crown's forces had provided minimal help—or worse, had actively resisted provincial defense efforts. Despite occasional assistance from individual Royal Navy guard captains, it would be New England's and Nova Scotia's provincial navies that played the largest role in defending Britain's northeastern maritime borderlands throughout the late 1710s–30s.

Provincial Navies and Imperial Borderlands: South Carolina, 1715–1727

Even before Massachusetts and Nova Scotia authorities went to war against pirates and Indigenous naval forces on the empire's northern borderlands, similar maritime violence erupted on imperial borderlands to the south and southwest of Charles Town, South Carolina, in 1715. Indigenous nations such as the Yamasee began to resent the South Carolina government's expansive goals, rumored plans of conquest, and abusive traders that enslaved (or threatened to enslave) Indigenous debtors.[53] By April of that year, disaffected Indigenous leaders killed two South Carolina traders and fired the proverbial first shots of the bloody Yamasee War. While the naval theater of the war that will be considered below primarily pitted South Carolinians against their Yamasee foes south of Charles Town, the colony also warred against other disaffected Native groups on land such as the Creeks and Choctaws.[54]

While disputes between southern colonists and their Indigenous neighbors were not as clearly tied to maritime matters as those in the Northeast, both South Carolinians and the Yamasee had strong ties to the sea. Natives living on the South Carolina coast had long engaged in maritime endeavors and were particularly skilled in crafting periaguas for trade, combat, and transport. In fact, Yamasee mariners had long used canoes in their role as agents of the ongoing Indian slave trade and, at times, even helped to man South Carolina's scout boat navy.[55] Colonial officials had designed this innovative fleet of two or more armed sailing periaguas in the latter years of Queen Anne's War—regularly keeping scout crews on the province's contested southern borderlands to patrol for Hispano-Indigenous incursions and to keep enslaved Africans from fleeing the colony's southernmost plantations.[56] In essence, the scout boat fleet was a small but semiregular coast guard that had the political and financial backing of South Carolina's government in both peacetime and war.

In yet another challenge to the prevailing notion that Indian wars were limited to land, the Yamasee nation and their Carolinian foes fought many of their battles on the coastal waters and streams near modern-day Beaufort and Port Royal, South Carolina. Both sides preferred small craft like their contemporaries in the Northeast but generally utilized shallow-draft periaguas and whaleboats over shallops and sloops. After Governor Charles Craven's militia forces stemmed the terrestrial Yamasee onslaught south of Charles Town in mid-April 1715, he directed the experienced frontiersmen

Alexander Mackay and John Barnwell to lead a naval assault against the Yamasee village of Pocotaligo. By the end of April, Barnwell and Mackay led militiamen on small craft to seize Pocotaligo and then seized a well-defended Yamasee fort after scaling its walls amid a hail of musketry.[57]

In September, South Carolina scout boat crews had conducted several successful ambushes against Yamasee warriors on canoes, including actions that involved coordinated land-based ambushes and musket/swivel gunfire from provincial vessels. The colony's scout boat navy was essentially purpose built for these campaigns on the colony's tidal borderlands. In historian Larry Ivers's view, the scout boat mariners had evolved from mere borderland scouts in Queen Anne's War to "marine commandos" by the end of 1715.[58] While this modern analysis may sound overly boastful of the provincial navy's progress, at least one contemporary South Carolina parson bragged in October 1715 that the worst of the crisis was over because "[as soon as a] party of Indians appear our Scouts give notice and they are beaten back."[59]

South Carolina's swift deployment of scout boats and militiamen on canoes was effective in stemming the initial Yamasee offensive, but these victories belied the large extent to which the colony depended on outside assistance from neighboring governments and imperial forces throughout the conflict. For instance, in May 1715, Craven's administration begged Governor Alexander Spotswood of Virginia for reinforcements and also asked Captain Samuel Mead of HMS *Success*—a passing Royal Navy warship—for supplies and to request direct help from London. While Mead refused both requests, he did agree to facilitate the purchase of weapons from Governor Joseph Dudley of Massachusetts.[60] Dudley agreed to the arms sale despite his own ongoing fights with Wabanaki mariners.[61] Even as New Englanders struggled against their own Native foes on the northern borderlands of British North America, they extended military aid to their compatriots facing similar issue across the continent.

Despite outside assistance and some tactical success against the Yamasees by late 1715, South Carolina's military situation was still dire. No account better describes the chaos than Captain Mead's December 1715 letter to the Lords of the Admiralty. Upon a subsequent trip to Charles Town, Mead reported that he was shocked to find the city bereft of defenders, with "the [governor] gone to the Army," and a great many men fleeing the colony to search for Spanish treasure after news of major shipwrecks on the Florida coast. With rumors that enslaved Africans planned to use the chaos to stage their own uprising, the lieutenant governor implored Mead to "send on Shore every night Twenty five, or thirty Men with Arms" to guard the city's powder magazine. The previously reticent Mead agreed to this plea.[62]

Mead's alarming description of the chaos in the proprietary capital would have been one of many flooding imperial offices in London in 1715 and 1716. Many South Carolinians themselves were beginning to resent the alleged military neglect of the privately owned colony's Lords Proprietors and their expectations that the colonists should orchestrate their own defense. In response, many provincial leaders began to campaign for direct royal governance and military aid. When the Board of Trade conducted a formal inquiry into the supposed neglect in the summer of 1715, the Lords Proprietors demonstrated a financial unwillingness to assist their colonists coupled with an outright ignorance of the colony's dire straits.[63] This ignorance and neglect was especially apparent when it came to the proprietary opinion regarding South Carolina's naval capabilities. In response to the Board of Trade's query as to whether the Lords Proprietors would provide shipping to carry British troops to the colony, they responded that "we do not doubt but the Governmt. of Carolina will send ships and provisions for their transportation."[64]

The Lords Proprietors' rosy view of South Carolina's ability to transport British regulars across the Atlantic clashed with the growing human and economic costs of the conflict. Because the Yamasees had found refuge with sympathetic Spanish authorities in St. Augustine, Florida, they continued to harass South Carolinians and even successfully ambushed one of the colony's scout boat crews near Port Royal in the summer of 1716.[65] That same year, South Carolina officials convinced the Cherokee—the traditional foes of the Yamasee nation's own Creek allies—to join the war effort. This alliance with one of the strongest southern Indigenous nations inspired several smaller neighboring tribes to join the English cause and helped Carolinian forces to bring the Yamasee to an uneasy stalemate.

This "stalemate" could hardly be called a victory; wartime losses included the deaths of over seven hundred colonists, food shortages, and nearly £116,000 sterling in war debt.[66] Continued skirmishes with Yamasee forces throughout the next decade would require constant vigilance, and the colony once more provided for "two scout boats of 10 men each" to guard the southern approaches to Charles Town.[67] These growing costs would play a signal role in the colony's decision to revolt against the parsimonious Lords Proprietors in 1719.

While the initial stages of the fight with the Yamasee nearly brought the colony to ruin, it would soon face an equally daunting threat: pirates.

Even though Anglo-American pirates did little to further cripple South Carolina's already-imperiled economy during the height of the Golden Age of Piracy, the threat of violent sea robbers off the coast did little to calm the anxious and war-weary populace. The political ramifications of piracy for Charles Town were most colorfully illustrated by Captain Edward "Blackbeard" Teach's June 1718 blockade of the unguarded port city and subsequent attacks by Teach's associate Charles Vane.[68]

Without any Royal Navy vessels nearby, Governor Robert Johnson, "thô very unable both for want of men and money," decided to commission militia officer Colonel William Rhett as a temporary vice admiral and authorized him to assemble an emergency fleet of pirate hunters. Johnson recorded that "two sloops [*Henry* and *Sea Nymph*], one commanded by Capt. [John] Masters and the other by Capt. [Fayrer] Hall with about, 130 men were gott ready wth. all the dispatch wee cou'd."[69] Even though the Royal Navy was far away from Charles Town at the time, Johnson clearly wanted Rhett's provincial fleet to carry the trappings of a Royal Navy squadron when he ordered Rhett to fly "his Majesties Union Flagg" on his vessels.[70]

Rhett's fleet never found Blackbeard or Vane, but they did capture the infamous "Gentleman Pirate" Stede Bonnet, in a pitched naval battle off the coast of Cape Fear, North Carolina, in September 1718.[71] The government's appropriation of Royal Navy trappings continued into the trial of Bonnet's crew. For instance, when one of the accused pirates made the nearly comedic claim that they only engaged Rhett's fleet because they thought they were being attacked by pirates themselves, South Carolina chief justice Nicholas Trott retorted: "And so one pirate might fight with another. But how could you think it was a Pirate, when he had King George's Colours?"[72] Even though provincial naval vessels were not part of the Royal Navy themselves, they adopted this exclusive royal banner to legitimize their pirate-hunting mission.

Aside from provincial compensation, sailors on pirate-hunting missions could also expect some level of reward from the home government in London. Thanks to King George I's 1717 promise of rewards for any sailors who captured unrepentant pirates, crews could expect financial gains up to £100 for the capture of a pirate captain and lesser amounts for lower officers.[73] The provincial government's burden of repaying its sailors was also lightened by the division of "booty" in vice admiralty hearings after Bonnet's capture. Almost simultaneously with the trial and execution in late 1718, Trott ensured that plunder from Bonnet's vessel was split equitably among the sailors.

Oddly enough, Trott required some of Bonnet's victims—merchant captains rescued by the South Carolina provincial navy—to pay salvage fees. This forced at least two "rescued" merchant captains to surrender their vessels to the South Carolina government due to their inability to afford these fees. During these proceedings, Trott also ruled that an enslaved African man named Ned Grant would be publicly auctioned off, and the proceeds of his sale would be used for prize money for the pirate hunters. Bonnet had captured Grant after he escaped from his South Carolina slaveholder and then had the misfortune of being recaptured by South Carolinians during the Battle of Cape Fear. Ironically, throughout these proceedings, the South Carolina government rewarded pirate hunters by seizing vessels from pirate victims and by depriving captives of their freedom.[74]

Even though the government at Charles Town was able to ensure that pirate hunters were adequately reimbursed for their services, larger piratical and political threats awaited the ever-embattled colony. With reports of new pirate fleets off the coast in the final months of 1718, Governor Johnson opted to expand his naval defenses and ordered a unique combination of scout boat patrols in the harbor and the impressment of an emergency fleet of merchant vessels to prevent the expected assault.[75] This would be one of the few times when South Carolina combined its regular naval forces with an impressed merchant fleet.

Worried about the damage that could come to their vessels after being impressed, several mariners complained to the provincial government and demanded assurances that they would be reimbursed for damages in battle. As financial negotiations continued, two pirate vessels under the command of Richard Worley appeared outside the harbor. Johnson dispatched the colony's scout boats to prevent their landing on the city's barrier islands and then led three hundred sailors on four ships and swiftly defeated the piratical duo. Johnson immediately split the "small booty" from the captures among the captors.[76] Once again, a timely deployment of an emergency fleet prevented Charles Town's foes from sacking the city.

While provincial victories over Bonnet and Worley raised Charlestonians' morale, increased tensions with the Lords Proprietors prevented any true respite. Many within the provincial government were infuriated with the proprietors for a number of reasons. High on the list of grievances were the facts that the Lords Proprietors vetoed the Assembly's military finance laws from the Yamasee War and that they failed to provide adequate military protection for their beleaguered and embattled colony.[77]

International politics would also play a role in heightening the crisis between the provincial government and its proprietary overlords. In response to Spanish attempts to expand their Mediterranean holdings in 1718, Britain and France had jointly declared war on the Iberian kingdom in the short-lived War of the Quadruple Alliance.[78] As the South Carolina militia gathered to prevent a rumored Spanish invasion, angry political activists formed an "Association" to discuss their dissatisfaction with the Lords Proprietors, fomented a bloodless coup, and installed a sympathetic governor that helped them to call for direct royal governance. This coup would forever be known in South Carolina as the Bloodless Revolution of 1719 or simply the Revolution of 1719.[79]

For at least some antiproprietary polemicists, the proprietary indifference that forced the colony to muster its own costly naval forces was a large justification for the Revolution of 1719. For instance, in a 1726 pamphlet that challenged proprietary attempts to retake the colony, South Carolinian Francis Yonge argued that one of the turning points that led to the coup was the proprietors' dismissal of a legislative session that had convened to find ways to settle military debts including the two missions against the pirates.[80] For men like Yonge, royal military protection was the only solution to their inadequate defenses. Even before King George I's privy council agreed to "provisionally" facilitate royal governance of the proprietary colony in August 1720, Captain John Hildesley of HMS *Flamborough* became the first Royal Navy post captain in Charles Town.[81] Seven years later, Yonge would praise the king for having "Protected . . . Trade by His Ships of War, and [the] Country by His forces."[82]

While Yonge was quick to praise Royal Naval protection in his antiproprietary pamphlet, he neglected to mention that South Carolina's provincial naval forces still continued to operate both alongside *and* independently of their new Royal Navy allies and that the presence of the Royal Navy did not guarantee internal stability. This was especially apparent during the period stretching from 1715 to 1732, which early twentieth-century historian Verner Crane called South Carolina's era of "Defense and Reconstruction." Crane argued that South Carolina expanded its southern frontier defenses during this era with numerous forts to challenge Franco-Spanish and Indigenous incursions, including the 1716 establishment of a more permanent base at Port Royal for South Carolina's two scout boats to operate out of.[83]

Despite provincial protestations in early 1720 that naval defense costs on the Florida frontier continued to exacerbate South Carolina's debt, the Board of Trade hinted at their desire for continued provincial naval efforts in September of that year. The Board of Trade expressed their desire that a fort should be built on the Altamaha River to the south of the colony (in what would become the colony of Georgia in the next decade) and were of the opinion that it would be difficult to do that without a Royal Navy guard ship. However, they proposed, in case a Royal Navy ship was not available, "that the [new Governor Francis Nicholson] be impower'd and have directions to hire a sloop or brigantine for this purpose upon his arrival in Carolina."[84] In essence, imperial authorities expected provincial authorities to fund their own naval defenses even as the royal military presence expanded in the region. As it would happen, the king's parsimonious Privy Council only provided a small unit of invalid redcoats for land defense, building materials, and a few officials to help with the fort's construction. Ultimately provincial authorities not only hired a merchant vessel and sailors to assist the expedition to build the fort but also used mariners from the scout boat navy to help build the fort itself.[85]

While South Carolina authorities continued to rely on their own mariners to secure imperial aims in the interwar period, they also came to realize that the presence of the Royal Navy could exacerbate internal political issues and instability. For instance, some Royal Navy guard captains enmeshed themselves in local corruption and political dramas. Early on in his tenure, provincial officials accused Captain Hildesley of *Flamborough* of partnering with provincial naval hero Rhett in an illegal arms cartel to the Spanish in St. Augustine. The controversial Royal Navy captain also conspired with ousted governor Johnson in an aborted attempt to retake his office in 1721.[86] If the Carolinians' allegations were true, Hildesley would have been better cut out for a modern mafioso than a Royal Navy station captain.

Provincial authorities would be forced to draw on their own naval resources again when imperial conflict with Spain temporarily resurfaced for a second short time in the late 1720s. Between 1727 and 1729, South Carolinians faced an onslaught of Spanish privateer raids on their merchant shipping as well as ongoing fights with Yamasee and Creek forces on the southern borderlands. In a letter from September 1727, one South Carolina merchant lamented that despite this combination of foes, heavy provincial taxation, and political uncertainty (i.e., the Lords Proprietors in London trying take back the colony) the citizens of Charles Town had "fitted out, at their own Expence, a Sloop [*Palmer*, captained by one Thomas Montjoy] with 100 Men" to hunt Spaniards alongside a Royal Navy guard ship, HMS *Scarborough*.[87] While other colonies with Royal Navy station ships occasionally fitted out their own provincial vessels, it is telling that South

Carolina—a colony that had orchestrated a political coup in part to secure royal naval protection—still felt the need to employ these vessels.

Historian Nic Butler has found that the South Carolina Council's decision to outfit that sloop coincided with the Royal Navy post captain George Anson's requests to impress local sailors for *Scarborough*. Butler contends that the overextended captain was already tasked with both hunting Spanish privateers and protecting merchant vessels and likely found comfort in the colonial government's decision to outfit a temporary "privateer" to defend the coast. The governor's council allowed the captain to impress sailors for his warship, even though contemporary British law (the 1708 America Act, which was also known as the Sixth of Anne) forbade Royal Navy impressment in the Americas. Despite this prohibition, the extent to which it forbade *all* impressment was unclear, and no imperial guidance existed to clarify when it was acceptable for colonial governors and Royal Navy captains to impress seamen. By the 1720s, after a decade of limited clarification over the law from Parliament, Admiralty officials had officially stopped requiring Royal Navy captains to avoid impressment—a unilateral decision that Parliament did little to challenge and that would have major ramifications in the decades to follow.[88] Not for the last time, a royal official depended on provincial naval forces to support the Royal Navy's meager presence in the area.

By late September 1727, the South Carolina legislature passed a comprehensive act that allocated funds for provincial naval and land forces to campaign against both Indigenous forces and Spanish privateers and that included a specific pay table for sailors aboard the *Palmer*, "Sloop of War Employ'd in Guarding These Coasts." This was to be funded by utilizing paper currency returned to the treasurer from a previous "Act for Calling in and Sinking the Paper Bills" and backing the expenditure with renewed duties on the slave trade and liquor sales.[89] This financial package of over £25,000 (in South Carolina currency) was especially impressive considering Middleton's and the Commons House's disagreements over expanding paper money then in circulation, as well as a recent royal prohibition on extending paper money circulation in the colony.[90]

Ultimately, Charles Town would wage a successful counteroffensive against Spanish and Indigenous forces throughout late 1727 and early 1728. In early January 1727/8, a Philadelphia newspaper reported that Mountjoy and the crew of *Palmer* retook a "ship belonging to London, which the Spaniards were carrying to the *Havana*; the ship and Goods was Praised and one Half was allowed for Salvage," though other Spanish privateers still cruised off the coast.[91] Soon thereafter, an Indigenous attack on a militia patrol boat south of Charles Town coupled with an insulting letter from Florida's Spanish governor convinced President (n.b. the title for an acting governor) Middleton to launch the long-planned assault on Yamasee lands near St. Augustine. Scout boat veteran John Palmer led nearly one hundred periagua-based English and Indigenous soldiers on a maritime expedition that devastated several Yamasee villages there.[92] This raid not only demoralized Spanish authorities in the area but convinced many Spanish-aligned Natives that their imperial ally could not adequately protect them.[93]

On the empire's contested northern and southern continental borderlands, imperial officials encouraged provincial officials to shore up imperial weaknesses with locally raised navies. While Anglo-Americans in Nova Scotia, New England, and South Carolina always pined for elusive Royal Navy assistance, they continued to pass legal and economic measures to ensure that provincial naval forces could fight simultaneous Native and piratical threats with or without royal assistance. Occasionally violent interactions between both provincial and Royal Navy forces foreshadowed larger rifts between imperial and provincial officials over the definitions and purposes of provincial navies in the next imperial war.

Provincial Navies, Privateers, *Guarda Costas*, and Pirates: The West Indies, 1713–1739

Between 1713 and 1739, West Indian authorities faced an entirely different interwar situation than their North American compatriots. For instance, despite some North American naval expansion, the Royal Navy continued to devote more ships and resources to the rich Caribbean islands than their continental neighbors. While the Royal Navy slowly expanded its operations in the West Indies during the previous two imperial wars, it ensured a more permanent presence in the West Indies during this with the construction of careening bases at Port Antonio, Jamaica, and English Harbour, Antigua, in the late 1720s.[94] By the early 1720s, there were nine Royal Navy guard vessels in the Caribbean, five stationed in the mainland North American colonies, and a couple of others stationed at the fishery in Newfoundland.[95]

Another major difference from the continent was that Anglo-American officials in the West Indies did not face major conflicts with powerful Indigenous tribes and primarily dealt with threats from Anglo-American pirates and Spanish *guarda costas*. It was also rare during this era of irregular conflict for Caribbean authorities to deploy land forces or use provincial vessels to transport troops to battlefields like their continental compatriots. Finally, Caribbean authorities relied on privateering far more than provincial officials on the mainland.[96] These differences between operational theaters can be overplayed however. Despite the regular presence of Royal Navy warships, those imperial commanders were not always willing or able to pursue the light craft of Spanish *guarda costas* or English pirates.

Just as with their compatriots on the mainland, Anglo-American officials in the Caribbean still often had to utilize their own naval resources to defend their shores and to secure British commerce between 1713 and 1739. Historian David Wilson's characterization of contemporary pirate expeditions as a "series of fragmented and distinctive campaigns, shaped and influenced in metropolitan and colonial contexts," succinctly describes the uneasy partnership between Royal Navy and provincial assets to combat not only pirates but *guarda costas* and regular Spanish military forces as well.[97]

This alliance must be understood in the context of inconsistent messages from the metropole regarding naval defense. The British government was both worried over the outbreak of guarda costa violence and Anglo-American piracy but also unwilling to expand its foreign military budget after the debts accrued in Queen Anne's War. Additionally, the Walpole administration was particularly hesitant to exert too much military on Spain as it wanted to create more trading opportunities. Walpole and his allies wanted to use the Royal Navy as a "passive force focused on deterrence." This metropolitan call for restraint coincided with tepid imperial support for local privateering initiatives in the colonies during this era; Walpole believed—perhaps with reason—that privateering was a gateway to piracy.[98]

However, as with their mainland North American colonies, imperial officials did not discourage provincial governments from exercising the right to raise their own defensive fleets as necessary. In March 1723, Barbadian governor Henry Worsley wrote to the Board of Trade about his fitting out a sloop "in the nature of a guarda costa" to assist the customs officer in pursuing smugglers. He wrote that he and the customs officer both had shares in the sloop and that the "expence will not be much to H.M. besides the maintaining the third part of the sloop."[99] In August, an unnamed imperial official (possibly Charles Delafaye, the undersecretary of state for the Northern Department) responded that Worsley's deployment of the sloop "employed in the Custom house affairs to prevent the running of goods, leave[s] no room to doubt but that step will be approved of," and doubted that the king would ask for any percentage of prizes taken in "so necessary a service."[100] With attitudes ranging from the apathetic to enthusiastic, imperial officials neither strongly supported nor discouraged provincial naval defense in this period. The burdens of coastal defense would almost always be the responsibility of provincial authorities.

Of course, any such laissez-faire approach to matters of colonial defense only worked when provincial authorities did not imperil Anglo-Spanish relations. Governor Lord Archibald Hamilton of Jamaica's controversial deployment of quasi-legal "privateers" in 1715 and 1716 would highlight just how such problems could arise. In 1715, Hamilton commissioned around ten privateers with the declared purpose of curtailing Spanish *guarda costa* and other pirate activity around Jamaica. Despite these declared aims, many of these privateers engaged in outright piracy against Franco-Spanish shipping. Far from having altruistic concerns for the safety of local mariners alone, Hamilton actually had large financial investments in these pseudo-legal voyages. Additionally, Hamilton was suspected of coordinating with Jacobite rebels against the Hanoverian dynasty, and it is possible that he may have intended this fleet to support the first large Stuart insurrection in 1715.[101]

Whatever his actual motivations, Hamilton defended his privateering campaigns by highlighting the inadequacies of Royal Navy support.[102] When questioned by Jamaican politicians in 1716 "what motive he had for granting ten Comissions in the Space of a month for Suppressing of Pyrates when a Kings Ship and a Sloop attended this Island." In response, Hamilton reckoned that the royal vessels had regularly been absent or in disrepair.[103] In a subsequent 1718 pamphlet, Hamilton clarified this answer and declared that local merchants reeling from *guarda costa* attacks had complained about the "want of Ships of War . . . [and desired that] such a Naval Strength may be order'd for the Protection of the Island."[104] Whatever Hamilton's personal dishonesty or ambitions, his complaint about inadequate Royal Navy support was far from groundless.

In a twist of historical irony, Hamilton's privateer fleet—ostensibly created for provincial naval security from Spanish "pirates"—helped to spearhead the Caribbean theater of the Anglo-American Golden Age of Piracy. Even though Hamilton had dictated strict parameters in his letters of marque—including requiring his privateer captains to fly a privateer jack rather than the Royal Navy Union Jack—some of his captains eventually turned to outright piracy against their own countrymen. This group of Anglo-American pirates (including the likes of Benjamin Hornigold and

Blackbeard) would eventually transform the Bahamian settlement of New Providence (now Nassau) into a major pirate base by 1716.[105] Whatever Hamilton's initial motives in commissioning his "privateer" fleet, his actions helped to spur on a wave of pirates that would force governors throughout the West Indies to expand their own provincial fleets to go pirate hunting.

By all accounts, Britain's immediate reaction to the growing pirate threat of 1716–18, ranging from general amnesties for penitent pirates to disorganized Royal Navy cruises, was inadequate. By 1718, Crown authorities finally supported a major campaign against the Bahamian pirate base because renowned privateer captain Woodes Rogers and private investors devised a colonization plan that required "minimal public expenditure."[106] Even though Royal Navy vessels helped Rogers to clear New Providence Island of pirates during his July landing, by September all of his Royal Navy escorts had departed.[107]

With few naval defense options remaining, the experienced privateer and new governor of the Bahamas utilized local naval strength in creative (albeit dangerous) ways. With threats from Spanish Cuba and the unrepentant English pirate Charles Vane, Governor Rogers resorted to hiring former pirates including Benjamin Hornigold and John Cockram to hunt down their old associates and ultimately met with some operational success. In a late October 1718 letter to the Board of Trade, Rogers—still bereft of Royal Naval aid and limited to help from local ex-pirates and his own crew of the private ship *Delicia*—suggested that any future Royal Navy vessels sent to the island should be under the direct command of the local government. With a small Royal Navy cruiser, Rogers "could joyne a sloop or two and men from the guarrison [sic] with the best of the people here and soon be out after any pirate."[108] Rogers's idea of a cooperative royal-provincial pirate-hunting force highlights what historian Mark Hanna has called "one of the first unified imperial projects."[109] Broadly speaking, it was a combination of provincial naval campaigns and Royal Navy cruises for pirates that helped to mitigate the worst of the Anglo-American pirate threats by the 1720s.[110] Though "unified" in cause against pirates and *guarda costas*, provincial and Royal naval forces were far from unified in mission or tactics in dealing with these threats.

The uneasy relationship between West Indian privateers, provincial navies, and the Royal Navy would continue throughout the next two decades as campaigns against pirates continued alongside a growing focus on Spain's imperial threat. No island better exemplifies occasionally unified but more frequently disparate provincial and royal paths taken against piratical and Spanish threats than Jamaica's experience from 1718 to 1729. In December 1718, a pirate captain captured a merchant captain with a lucrative cargo, and Governor Sir Nathaniel Lawes dispatched two provincial crews in pursuit as "none of H.M. ships of war [were] then in harbour." Lawes promised the sailors one-third of the shares of "whatever was recovered" as delineated by a recent royal proclamation against pirates. The crews found the culprit—a Spanish pirate with a multiethnic crew—and fell back in defeat after a bloody engagement with over thirty-five sailors killed and more wounded.[111]

Soon after this defeat, Lawes informed his council that "Several Merchants had Voluntarily offered their Sloops Tackle and ffurniture and to fit them out on the Credit of the Country towards further pursuing the Pyrates." The council agreed that "three Sloops should be forthwith sent out ... be Arm'd and Victualled by the Government" and that the "Sloops be at the Risque of the Owners, and Mens Wages as the others had been before." The council appointed specific captains, demanded they "Concert together and have the same Commissions and Instructions as the Two former." What the council suggested was an interesting combination of a governmentally directed naval assignment and a privateering mission guided by the ethos of "risk and reward."[112]

While the new provincial fleet was on its hunt, six Royal Navy warships arrived in Kingston Harbor. Far from praising the arrival of royal military aid, Lawes complained to the Board of Trade that he had little control over the warships' captains. In one case, Lawes—still ignorant of the recent declaration of war between Great Britain and Spain—complained that a Royal Navy captain failed to deliver his letter to the governor of Cuba inquiring about Spanish attacks on Jamaican shipping.[113] Luckily for Lawes, those sloops "fitted out at the charge of the country in pursuit of the pirate that took the ship *Kingston*, are return'd with pretty good success" and recovered the vessel without firing a shot. Around the same time, Lawes reported that Jamaican "privateers have already made application for Commissions to act against the Spaniards, and I have with the advice of the Council issued some."[114]

Throughout the rest of the short-lived War of the Quadruple Alliance and beyond, Lawes and his successors struggled to harmonize Royal Navy and provincial maritime defense strategies and goals. This became immediately obvious in the Jamaica government's decision to regularly employ guard sloops despite a Royal Navy presence throughout the 1720s.[115] Desperate to fund regular provincial guard sloops while also cutting costs, Lawes informed the Board of Trade in late 1720 that he had agreed to the

Jamaican Assembly's levies on slave sales and proposal to tax Jewish residents £1,000 in order to fit two vessels for the "guarding the coasts from pirates and other vessels from Trinidado [Cuba] who frequently commit depradations and acts of hostility both by sea and land upon us."[116]

While the politics surrounding Jamaica's expansion of its provincial navy could lead to discriminatory treatment of its minority populations, they could also alienate Royal Navy officers like Admiral Edward Vernon. Tensions were already high between Vernon and Lawes's administration because Vernon consistently involved himself in local affairs—ranging from attempting to extradite alleged Jacobites to England for trial to accusing Lawes and his attorney general of smuggling.[117] In late 1720, in a preface for an act to fit out guard sloops, the Jamaican Assembly made the following barb: "Whereas ... H.M. ships of war ordered here for the encouragement of trade and defence of this Island have not so effectually answered the end for which they were sent hither ... whereby a great many ... vessels as well belonging to this his said Island ... have been taken in sight thereof by pirates and vessels fitted out and commissioned by the subjects of the King of Spain under pretence of guarding their own coast." Vernon responded to this accusation in a letter to the Admiralty, claiming that Royal Navy ships were not equipped to chase after Spanish-sponsored pirates in small craft and that the Assembly's accusation was a "lying preamble."[118] Just as with the arrival of the Royal Navy in South Carolina in the 1720s, Jamaicans came to realize that their coastal defense could be left to imperial authorities alone. By April 1721, Lawes was confident enough to brag to London that "I am told our adjacent Spanish Governors are grown more cautious in granting commissions to guard de la coasts especially since the country sloops have been cruiseing round about the Island."[119]

While provincial forces could be useful substitutes for lackluster Royal Naval forces, it is important to note that the situation was nuanced and that there were times where Royal Navy captains were proactive in defending provincial interests. For instance, In February 1724/5, the next governor—Henry Bentinck, the Duke of Portland—disagreed with the Jamaica Assembly over various legislation and attributed the failure of a new coastal security bill to this dispute. The governor bragged that "I have not sufferd your Coast to lye Naked, The Commadore having at my instance (very readily indeed) commanded his Majesties Snow to that Station."[120] The duke reported to the Board of Trade that "I prevaild wth. the Commadore to order one of H.M. sloops to supply the want of the guard sloop," that the Assembly accepted this, and hoped to use the initial money for a provincial guard sloop to suppress an ongoing slave rebellion.[121] At least in this unique case, the Royal Navy was more dependable in terms of coastal defense than their provincial counterparts.

These disparate views over the best course of naval protection became even more evident in 1726. An anonymous pamphleteer appealed to the royal government for permission to enact reprisals against Spanish *guarda costas* for shipping losses and later decried the alleged inactivity of large Royal Navy warships to pursue swift foes, as well as the inability of the island to keep financing expensive provincial guard sloops. The governor did admit, however, various political disagreements with the Lords of Admiralty as well as the "negligent and disrespectfull behaviour of most of the Sea Officers" in the area.[122] The pamphleteer and the governor may have disagreed over the Royal Navy's success in hunting *guarda costas*, but they both highlighted major problems that threatened the province's naval security: heavy financial costs for provincial officials in outfitting local defense fleets and personal disputes between local officials and Royal Navy officers.

Notwithstanding the lack of cohesion between provincial forces and the Royal Navy, both Jamaicans and the Crown continued to rely on this uneasy balance of provincial and royal naval action when a second (though largely uneventful) imperial conflict broke out again with Spain in 1726. While its causes were imperial disputes in Europe, local tensions in the West Indies—particularly an uptick in guarda costa attacks—provided a tense American background for this renewed war.[123] From 1726 to 1727, Vice Admiral Francis Hosier led an infamous attempt to blockade the Spanish treasure fleet in Porto Bello, Panama. Over three thousand Royal Navy sailors, including Hosier himself, would die from an epidemic of yellow fever during the failed blockade.[124]

Jamaican participation in Hosier's campaign was limited, but when King George II sent a letter in the spring of 1729 warning that Spain had plans to invade the island, local authorities scrambled to ready both the Royal Navy and to expand local maritime defenses. The Jamaican government's naval defense plan proved that provincial and Royal Navy vessels could operate in tandem on a large scale if needed. In early April, Commodore St. Loe promised Jamaican governor Robert Hunter that he "shall be ready to come into any Measure with Your Excellency for its safety, and shall keep those Ships that are with me in readiness to go upon any Service."

While the commodore's ships were cruising for enemy vessels, Hunter devised a plan for the defense of Kingston Harbor that relied almost entirely on merchant vessels. He proposed quickly training sailors from each ship to operate artillery at the fort, and that various merchant vessels should be armed, reinforced, and strategically placed to prevent landings. In the case of a successful Spanish incursion, the merchant captains were to fall back to Kingston and land their men. The merchant captains were to keep a "strict discipline amongst their People according to the Law of Arms," and financial insurance would be provided for wounded sailors and the families of sailors killed in action. His council agreed to these proposals and only added that sailors on land would be put under the command of the local militia.[125] While no Spanish invasion would reach Jamaica's shores, Hunter's simultaneous reliance on Royal Navy cruisers for external scouting and merchant vessels for emergency defense exemplified the ways in which a royal-provincial naval alliance could work on the field.

Ironically, the threat of Spanish *guarda costa* violence would be at its height during the "peacetime" following the short-lived Anglo-Spanish War of 1727–29. Historian Richard Harding has found that of the seventy-seven British vessels taken by *guarda costas* between 1713 and 1731, 34 percent of these vessels were taken after 1727. Negotiations in the early 1730s between Spain and Britain to end both English smuggling and Spanish *guarda costa* activity failed, and Spain and France renewed their ancient alliance in 1733—a worrisome prospect for the Walpole administration.[126]

Just as tensions were heating up with Spain, Rear Admiral Charles Stewart agreed to dispatch Royal Navy warships to seek restitution for Jamaica ships that had been taken by *guarda costas* in the autumn of 1732.[127] Even this newly proactive stance did not eliminate the need for provincial naval activity. A few months after this decision, a Spanish vessel seized an English sloop in the harbor of Port Morant, Jamaica, and "Two Sloops were immediately order'd to go in quest of the said Pirate."[128] Though the master of the sloop was released by the Spanish and restitution made, Anglo-American authorities had still felt it necessary to send sloops out to look for the missing merchant.[129]

Despite growing Royal Naval involvement in the region by the early 1730s, provincial naval activity continued in many areas. In early 1735, a Spanish guarda costa, with over 120 hands "most Negroes and Mollattoes" with two prize sloops in tow, attacked an English sloop and forced it to run ashore on the Isle of Saba. His "Excellency General Matthew . . . instantly fitted out a Sloop of his own with sixty Men to go in quest of the Pirate and her prizes." Even though he could not find the *guarda costa*, the "Example was followed by another gentleman who fitted out a Sloop at his own Expence" but did not have enough sailors himself. With the *guarda costa* still at large, "till better Measures can be thought of, out of his great Generosity, is fitting out his Sloop a second time for the Security of the homeward-bound Ships."[130] Even at this late date, merely four years before the next great imperial contest—the War of Jenkins' Ear—West Indian officials still felt the need to fit out small provincial vessels "till better Measures"—namely adequate royal naval assistance—could be thought of.

From the end of Queen Anne's War in 1713 to the beginning of the War of Jenkins' Ear in 1739, Anglo-American officials on Britain's North American borderlands and in the West Indies navigated a maelstrom of piratical, Native, and imperial maritime threats by developing flexible systems of provincial naval defense. Though occasionally deploying privateers, colonial officials on the contested borderlands of New England, Nova Scotia, and South Carolina depended on centrally controlled emergency fleets and guard vessels to pursue Wabanaki, Yamasee, and piratical raiders. While occasionally outfitting guard vessels like their continental brethren, West Indian officials dispatched impressed vessels, privateers, and *guarda costas* and pirates. Though imperial officials slowly expanded the Royal Navy's physical presence throughout the Atlantic world in the interwar period, parsimony, operational difficulties, and a guiding ethos of military restraint ensured that provincial naval forces would be necessary to secure the empire's maritime security in this interwar period.

Frontispiece to Captain George St. Lo's *England's Safety: Or a Bridle to the French King* (London: W. Miller, 1693). A late seventeenth-century depiction of Royal Navy sailors. It is likely that provincial sailors were dressed similarly during this era.

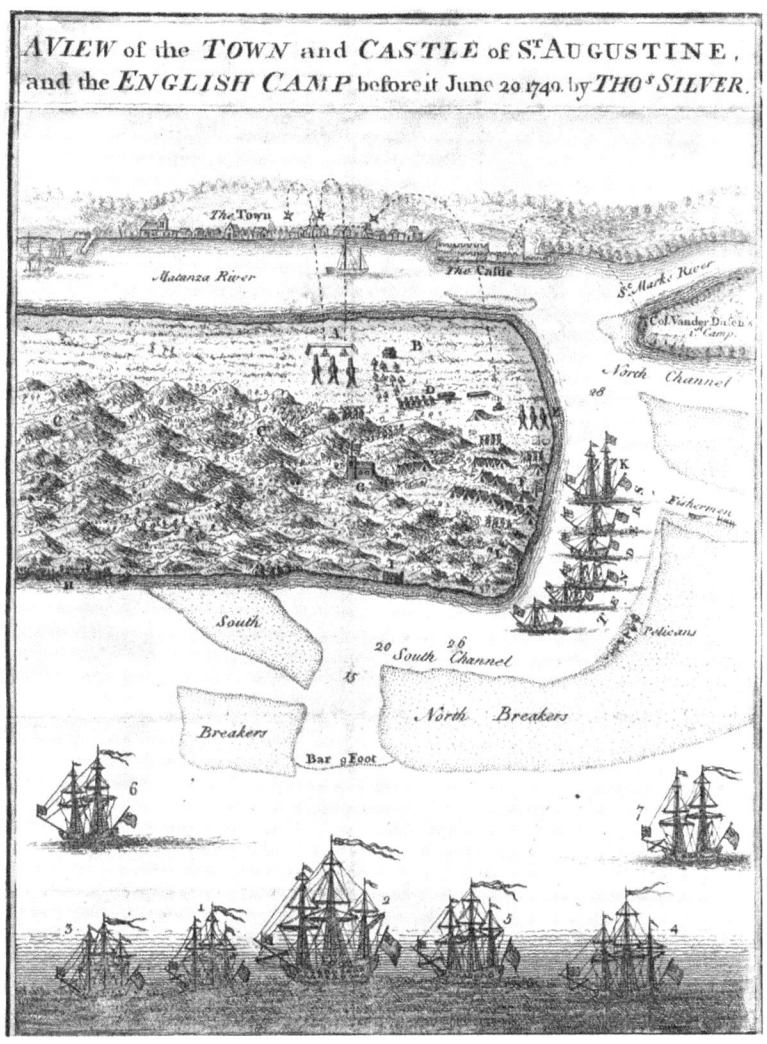

Thomas Silver, *A View of the Town and Castle of St. Augustine, and the English Camp before it June 20, 1740*, *Gentleman's Magazine*, November 1740. A British view of the failed joint Royal Navy–provincial attack on St. Augustine, Florida, c. 1740. The larger ships at the bottom of the diagram are Royal Navy warships, while those "tenders" in the foreground are likely the provincial warships under their command.

1740s engraving of the provincial frigate *Massachusetts*.

Sir Peter Warren by Thomas Hudson. Vice Admiral Warren was among the most active Royal Navy commanders in North America during the War of Jenkins' Ear and was an avid advocate for colonial provincial navies. Warren appears in the first generation of "navy blue" uniforms that would become iconic in the service. Wikimedia Commons.

British Resentment or the French fairly Coopt at Louisbourg by Louise Pierre Boitard. This scene from a larger 1750s political cartoon depicts "A Gang of Brave Saylors" (presumably from a Royal Navy warship) mocking their French foes who were "Coopt [up] at Louisbourg." While a large provincial fleet worked with the Royal Navy to capture Fortress Louisbourg in 1745, the British government returned the fortress to the French during peace talks in 1748. By 1758, a mostly royal force would recapture the fort for the British Empire. A special thanks to Kyle Dalton of the *British Tars* material culture blog for first leading me to this image.

Philip Dawe, *The Bostonians Paying the Excise-man, or Tarring and Feathering*. Printed for Robert Sayer and John Bennet (London, 1774). This political cartoon from 1774 depicts the active involvement of mariners in the violent protests and turmoil immediately preceding the Revolutionary War. The sailor on the right of the image is dressed in the manner typical of late mid- to late eighteenth-century seamen with a monmouth cap, petticoat breeches, short jacket, and neckerchief. Wikimedia Commons.

Frigate South Carolina by Jonathan Phippen, c. 1793. The South Carolina state Navy frigate *South Carolina* saw limited service in the latter end of the American Revolution before its capture by the Royal Navy in 1782. John Joyner, its captain, had previously served as a provincial scout boat captain in South Carolina. Joyner's career is a material reminder of the connection between provincial and later Revolutionary naval service. Wikimedia Commons.

4

The War of Jenkins' Ear and the Incomplete "Royalization" of American Naval Defense

When Great Britain declared war on Spain in 1739, few imperial officials could have expected a nine-year conflict that would ultimately pit them against the French as well. This conflict, the War of Jenkins' Ear—which would later bleed into a larger imperial and European conflict called the War of the Austrian Succession/King George's War—was different from the wars against Louis XIV in that it was largely inspired by maritime tensions.[1] By the late 1730s, after a series of failed negotiations over British navigation rights and Spanish *guarda costa* activity, Prime Minister Robert Walpole's political foes convinced his ministry to declare war on the Spanish.[2]

This fight against the Spanish had roots in navigation disputes in the West Indies but also in territorial disputes on the North American mainland that directly involved provincial naval forces. With borderland conflicts common between South Carolina, Britain's southernmost colony, and the Spanish and their Native allies throughout the last century, British statesman James Oglethorpe and his allies in the government planned a "buffer" colony for the region south of South Carolina called Georgia. Pleased with this arrangement, South Carolina's legislature immediately sent the scout boat *Carolina* with ten sailors to protect Georgia's earliest settlers in early 1733. *Carolina*—along with several other small provincial vessels—would prove to be vital in Oglethorpe's provocative military expansion south of Savannah at Frederica, Georgia, in 1736.[3]

British expansion in Georgia, backed by provincial war vessels, enraged Spanish authorities at St. Augustine and created the potential to blow up into all-out war. This border dispute took on greater imperial dimensions when Oglethorpe and his Spanish counterpart, Governor Francisco de Moral y Sánchez, agreed to cease expansion and let their respective overlords in Europe decide the Florida-Georgia line. The British secretary of state/Duke of Newcastle, Thomas Pelham-Holles, took a hard line against rumored Spanish military plans to invade Georgia and promised to protect the fledgling colony. He also secured Oglethorpe a place as the overall military commander of both South Carolina and Georgia. When Spain demanded a complete British withdrawal from Georgia, prowar politicians—along with King George II—agreed to send Royal Navy reinforcements and British troops to the New World.[4] It is telling that British military expansion in Georgia—largely contingent on provincial naval assistance—would ultimately help propel the British Empire to war against Spain in 1739.

Any study of provincial navies in the War of Jenkins' Ear must consider the massive expansion of the Royal Navy's war-making capabilities in this period. By all accounts, the Royal Navy would achieve "naval supremacy" by the end of the war in 1748.[5] When the Newcastle ministry made the final push for war in 1739, the Royal Navy had over 117 vessels in serviceable or nearly serviceable condition, twenty of which were stationed in North America and the West Indies. In terms of finance, Walpole's reduction of the national debt in the interwar period coupled with a decades-old banking and finance system allowed Britain to bankroll the war effort.[6] Logistical improvements also occurred throughout the war, including the direct Admiralty takeover of victualling for Royal Navy ships in Jamaica—a task that had previously been handled by inefficient private merchants.[7]

The Admiralty also made leaps toward greater professionalization during the nine-year conflict, although its transformation of the navy into a more regimented and uniform fighting force would take years to complete. In late 1744, when the John Russell, the Duke of Bedford became the first Lord of the Admiralty, he brought with him a host of new Admiralty Lords (the naval hero George Anson, John Montagu, the Earl of Sandwich, and George Grenville) that historian Richard Harding has described as a "generation of politician administrators." Between the mid-1740s and early 1750s, these new bureaucrats successfully cemented the Admiralty at the helm of Britain's disorganized naval governance web, instituted stricter discipline for seamen and officers, advocated for more offensive naval warfare (particularly in North America), ordered officers to wear uniforms for the first time, and ordered the construction of more powerful warships.[8]

Coinciding with the Royal Navy's administrative reforms and expansion in the War of Jenkins' Ear was the massive growth of privateering throughout the Atlantic world. In fact, there were more privateers in the War of Jenkins' Ear than in any previous conflict. Even though Britain encouraged privateering, it struggled to codify and enforce new regulations on the rapidly growing practice.[9] While privateering expanded during this conflict more than ever, so too did the proliferation of colony-funded provincial navies. Even though borderland colonies had funded some small provincial forces in the interwar period, the war prompted most Anglo-American provinces from Nova Scotia to Barbados to fund guard vessels and local navies on a much larger scale than ever before. Dozens of small, medium, and large provincial warships (e.g., the South Carolina galley *Charles Town*, the Massachusetts frigate *Massachusetts*, and the Rhode Island sloop *Tartar*) patrolled shipping routes, intercepted enemy privateers, and spearheaded naval assaults on enemy ports such as Fortress Louisbourg and St. Augustine.

In essence, provincial naval forces returned to the offensive and defensive tasks they had taken up in Queen Anne's War but on a much larger scale. For instance, for the first time, Massachusetts provincial vessels hunted for Spanish prey as far south as the Carolinas, and Rhode Island's guard sloop transported its colony's troops to the West Indies for operations in Cuba.[10]

The expansion of naval and privateering forces was part of an even larger midcentury expansion of British shipping around the Atlantic world. By the 1740s, regularized ship traffic along with mail and newspaper circulation increased in the English Atlantic world and transformed the British Empire "from a rimless wheel of dissimilar trades into a linked community." In this growing Atlantic web of commerce and travel, Anglo-Americans also became more aware of other colonies' provincial naval efforts and shared experiences of coastal defense. Take for instance a 1743 South Carolina newspaper report that "The *Boston* Province Snow (*Prince of Orange* [italics mine]) commanded by Capt. [Edward] Tyng, was spoke with on Wednesday last, cruizing off our Bar." Where provincial navies had operated on mostly regional terms before, by the 1740s, colonial vessels had begun to patrol waters far beyond their regional homeports.[11]

Growing political interconnectedness throughout the British Atlantic world had another side effect: by the second quarter of the eighteenth century, Anglo-Americans had come to see themselves as indispensable members of an increasingly interconnected British Empire. Scholars have long recognized this era as a point when competing provincial and metropolitan visions of empire began to diverge in dramatic ways. On the one side, Anglo-Americans had largely come to expect royal military assistance while still insisting upon political autonomy. On the other side, British officials had become more willing to provide direct military assistance to provincial governments yet were also more interested in tightening metropolitan control over those peripheries.[12]

These diverging views of the place of the colonies within the sphere of the British imperial constitution would have severe ramifications in the joint royal-provincial maritime campaigns of the War of Jenkins' Ear. Throughout the conflict, the British government began to intervene in colonial naval affairs more proactively by bankrolling various provincial navies and by more proactively deploying Royal Navy squadrons to assist provincial forces on the northeastern borderlands of North America. Nevertheless, vagueness in parliamentary legislation regarding prize distribution coupled with perceived misconduct by Royal Navy officers not only angered Anglo-American officials but also set the stage for future conflicts between Anglo-Americans and Crown forces at sea.[13]

"WHERE THE KING'S SHIPS ARE NOT": PROVINCIAL NAVAL WARFARE IN THE SOUTHERN COLONIES AND INCONSTANT IMPERIAL INVOLVEMENT, 1739–1744

Maritime tensions were the main causes of the War of Jenkins' Ear. Nevertheless, throughout much of the first half of the conflict, Royal Navy vessels failed to adequately protect the North American shoreline from Spanish privateers—prompting nearly universal adoption of provincial guardships and growing awareness among Anglo-Americans of their own (and their neighbors') naval power. After paying scant attention to colonial naval defense concerns for the first few years of the conflict, outcries over Royal Navy negligence on the Carolina station coupled with General Oglethorpe's campaign for financial compensation for outfitting a provincial navy inspired Parliament to support provincial naval forces for the first major time.

In an April 1740 report to the Board of Trade, royal customs surveyor and administrator, Robert Dinwiddie, estimated that British colonists from Canada to the West Indies operated over two thousand seagoing vessels, while British vessels traveling to the colonies numbered around one thousand. Dinwiddie guessed there were around 24,680 sailors operating out of Britain's Atlantic colonies at that time.[14] Whatever the accuracy of this report, by the beginning of the War of Jenkins' Ear, imperial officials—and Britons at large—were becoming increasingly aware of the scale of growing provincial naval capabilities.[15]

The provinces' naval capabilities were especially evident to Royal Navy officials. By April 1745, Admiral Peter Warren—a Royal Navy commander who had previously worked alongside provincial fleets in the failed attack on St. Augustine in 1740 and who would soon help to lead them in the siege of Louisbourg—wrote that imperial authorities should encourage every colony to fit out their own provincial navies that would be on the "same footing" as royal ships.[16] Problematically, imperial officials never created such a standard policy based on Warren's suggestion, and provincial officials continued their forebears' policy of funding local navies on their own.

In fact, Anglo-Americans not only took the initiative to defend their coasts from growing numbers of Spanish privateers but began to appreciate and study the naval efforts of their neighbors for the first time.[17] For instance, in Quaker-dominated Pennsylvania, various executives attempted to use news of other colonies' naval forces to pressure pacifistic assemblymen into raising a Pennsylvania navy. In 1741, Lieutenant Governor George Thomas complained to the assembly that it would be "very disreputable to this Province . . . to remain inactive, When Boston, Rhode Island, & New York, are fitting out Vessels of fforce to secure their Navigation by attacking the Enemy."

Seven years later, his temporary successor, Anthony Palmer, remarked that Pennsylvania should fit out a "Ship-of-War" to assist a Royal Navy sloop in the area. After all, the "neighbouring Colonies of New England, New York, Virginia, South Carolina, or the West India Islands . . . have almost all at times found it necessary, notwithstanding the Guardships station'd among them, to fit out Vessels of War to act in conjunction with those Guardships, or independant of them as Circumstances required." Not only were those vessels useful, "being immediately under the Command of their respective Governments . . . obliged to Cruize . . . where . . . the King's Ships are not," but they were also signs that a colony was not unwilling to "do all in its own Power" to assist royal military efforts.[18]

While Pennsylvania's government saw the necessity of local naval forces with constant harassment of enemy privateers, the southern governments of South Carolina and Georgia regularly relied on provincial naval forces to defend themselves from invasions—and to augment the meager Royal Navy presence in that region.[19] Even before war broke out in 1739, General Oglethorpe had been using four Georgia and South Carolina scout boats to assist the resident Royal Navy sloop *Hawk* in various operations around the coast.[20]

Unfortunately for Oglethorpe, paying for this substantial provincial fleet was more complicated in Georgia than in colonies with governors and assemblies. Georgia was technically a private colony run by a board of trustees in London, but these trustees depended on substantial yearly parliamentary grants for the colony's civil and military maintenance. By 1738, Walpole promised the trustees that the imperial government would cover the colony's military costs if the trustees continued to apply to Parliament for grants for nonmilitary costs.[21] Despite this assurance, no immediate funds came from London for local forces, and Oglethorpe ended up personally funding his provincial navy for the first few years of the conflict while holding out hope that Parliament would eventually reimburse him.[22]

In the tense autumn of 1739, Oglethorpe received King George's military instructions: to "make an Attempt upon the Spanish Settlement at St. Augustine" if he could get the cooperation of the South Carolina government and to encourage privateering against Spanish shipping.[23] While Oglethorpe would ultimately lead a major siege of the Spanish city with naval and infantry assistance from both South Carolina and the Royal Navy throughout the first half of 1740, the expedition—like the contemporaneous attack on Cartagena in Colombia—was an utter failure. In the immediate months and years following the defeat, contemporaries and historians alike have argued over who was most responsible for the campaign's failure.

From the very beginning of the joint campaign, tensions were high. After a successful raid on Spanish forts north of St. Augustine in late winter, a confident Oglethorpe implored South Carolina to assist him with the capture of the Spanish capital. South Carolina's government was slow to respond to Oglethorpe's and Royal Navy commodore Vincent Pearce's requests for aid, notably refusing to provide engineers or a deadline for when they would be ready. Even when the South Carolina legislature agreed to aid their compatriots, they still required an immediate loan from Oglethorpe in pounds sterling to outfit their troops and vessels.[24]

Despite this haggling, Benjamin Franklin's *Pennsylvania Gazette* reported in May that by early April "the [South Carolina] General Assembly have empower'd . . . the Lieutenant Governor to raise a Regiment of Foot, and a Troop of Rangers to assist General Oglethorpe . . . in Conjunction with several of his Majesty's Ships of War: as also to provide Sloops, Boats, Guns," and other necessities.[25] Indeed, contemporary estimates of both colonies' provincial fleets during the expedition highlight the scale of this undertaking. Around that same time, Oglethorpe reported that he employed three sloops, a long boat, a schooner, and numerous armed small boats including a "Colony Periagua being a Guard De Coast." This fleet included over 140 sailors and officers and cost Oglethorpe an extraordinary sum of £453 a month.[26] South Carolina's government provided an armed schooner with fifty-four "Volunteers and their [enslaved] Negroes" and numerous "Craft, Viz: 3 Sloops, one of which attended the [Royal Navy] Men of War," with twenty sailors in total, and "14 Schooners and Decked Boats," which employed over fifty-six armed men and sailors.[27]

While the *Pennsylvania Gazette* bragged that the naval efforts were done in "conjunction" with the Royal Navy, provincial officials largely placed their naval forces under the command of Commodore Pearce—captain of HMS *Flamborough* and at least eight other frigates and smaller Royal Navy vessels.[28] Pearce's relationship with his colonial compatriots was more troubled than almost any previous station captain before him, and would serve as a harbinger for the increasing wartime tensions between Royal Navy officers and provincial authorities.

One major operational pressure was the ever-present provincial anger over Royal Navy impressment policies. Even though Parliament had banned impressment in American colonies in 1708 with the Sixth of Anne, Admiralty officials stopped requiring personnel-depleted Royal Navy station captains to follow this act by the mid-1720s. These impressments continued throughout the war, and Parliament gave them official sanction in 1746 when it condoned impressment in mainland North America but banned it in the more lucrative West Indies colonies—a double standard that infuriated many Anglo-Americans.[29]

While colonial governments did still periodically opt to "Impress a Sufficient number of men and make provision for their subsistence" for service on provincial navy vessels, Royal Navy impressment was far more frequent and unpopular with Anglo-Americans.[30] During the preparations for the siege of St. Augustine, a privateer named Captain Davis sailed south to Tybee Island, Georgia, after "his Men [were] impressed into the Men of War, and himself engaged in much Controversy at Law . . . [, which] put a full End now to any farther Thoughts about privateering." He chose to "admit the Sloop he had with him into the publick Service, among so many others employed" by Oglethorpe. Davis's decision to join Oglethorpe's provincial navy not only hints at the possibility that provincial service could be an occasional escape from Royal Navy impressment but also highlights the malleability between independent privateering and provincial navies.[31]

In a more extreme case, the *South Carolina Gazette* claimed that in May 1740, a press gang from HMS *Tartar* tried to requisition several sailors from the merchant ship *Caesar* in Charles Town Harbor, and in an ensuing scuffle, royal sailors killed one of the merchant men.[32] Governor William Bull wanted to have the man responsible for the killing put on trial, but the Royal Navy captain and his crew set sail for St. Augustine before proceedings could begin. This evasion of the law, argues military historian James P. Herson, "made for bad press and may have contributed to poor contemporary and historical hindsight" regarding the siege—and particularly regarding Pearce and his Royal Navy squadron.[33]

While impressment surely soured some contemporary and future scholarly opinions of Pearce's leadership, it is worth noting that there were moments of cooperation during the preparations for the siege. For instance, when in April 1740 the crew of HMS *Squirrel* had captured a Spanish schooner, Captain Peter Warren asked the South Carolina government to operate the schooner with ten local sailors as an "Advice Boat to bring any Intelligence . . . for the Service of this Government."[34] All in all, it seems Royal Navy officers were more than happy to utilize South Carolina's provincial naval resources—whether through unpopular methods of impressment or more routine requests for provincial vessels to assist Royal Navy ships.

While provincial and royal forces faced many challenges throughout the several weeks of the siege, one of their primary obstacles was the presence of six well-armed Spanish half-galleys in St. Augustine's Matanzas Bay. These dangerous vessels and complement of two hundred sailors evaded British warships and gave the Spanish garrison total command of Matanzas Bay. By mid-June, Royal Navy officers' fears of the impending hurricane season combined with sundry mishaps including the Spanish defeat of the English garrison at nearby Fort Mose left provincial leaders desperate for a quick solution to taking the well-fortified town.[35] Throughout much of the second half of June, Colonel Alexander Vanderdussen—overall commander of South Carolina's troops—continuously offered to spearhead a joint provincial–Royal Navy assault on the Spanish half-galleys. Despite his persistence and support from Royal Navy captain Warren, Commodore Pearce considered the plot too dangerous to undertake. Within days, storms forced the warships out to sea, Cuban resupply vessels once more made it to the St. Augustine garrison, and the siege was all but over.[36]

Contemporaries and historians have long debated whether provincial forces or the Royal Navy was more culpable for the ignominious withdrawal from the siege in early July. Historian Trevor Reese makes the convincing case that imperial support for the expedition was limited as most Royal Navy forces in the Americas were concerned with the concurrent siege of the South American port of Cartagena. Reese contends that imperial preference for control over the more lucrative Spanish West Indies prevailed over provincial North American desires for a strong assault on St. Augustine.[37]

Whatever fault imperial authorities had for the failure, many of the leading figures in the siege blamed each other for the defeat. Between 1740 and 1743, South Carolinians, British army officers, and Oglethorpe's allies engaged in a vicious pamphlet war over the controversy.[38] Indeed, in one example, a South Carolina legislative committee complained that Commodore Pearce was "always declaring himself ready to give any assistance but never giving any at all" and was unreasonable in his dismissal of Vanderdussen's plan of attack. Perhaps most damning, the commodore had left South Carolina's "Province Schooner" (presumably the *Pearl*) to "shift for herself" at the mouth of Matanzas Bay without any assistance from the Royal Navy.[39]

With little time to recuperate from the defeat at St. Augustine in 1740, Lieutenant Governor Bull in South Carolina and General Oglethorpe in Georgia both had to adapt their provincial naval establishments to increasing waves of privateer attacks against the two colonies' shipping. This wave of enemy privateering between 1740 and 1742 covered the entire east coast of the British colonies but was particularly dispiriting for Carolinians and Georgians who had just faced a humiliating repulse. Historian Carl Swanson estimates that by the end of the war in 1748, Spanish and French privateers would capture 736 English vessels in North America and the West Indies. Despite the occasional capture of a Franco-Spanish privateer, this swarm of enemy vessels throughout the Atlantic—like so many bees—was too expansive for Royal Navy station ships or provincial navies to adequately handle.[40]

Despite the Admiralty Board's October 1740 instructions for Royal Navy ships to expand their patrols off the South Carolina and Georgia coasts, their presence was inadequate to defend local commerce. Provincial anger at alleged Royal Navy indolence would inspire repeated provincial petitions for greater naval assistance from London.[41] This is evident in Bull's October 1741 letter to the Duke of Newcastle, wherein he decried "the Interruption [Spanish privateers] give to the Trade of this Province; more especially at this time when his Majesty's Ship *Phoenix* is unfit for service ... [and] his Majesty's Ship *Tartar*" was due to leave soon—a fact that would leave the province defenseless and ripe for raiding. With privateers frequently fleeing to the refuge of shallow waters where Royal Navy frigates could not pursue them, station ships were essentially useless—a fact that inspired Bull to plead with the Duke of Newcastle to send material and laborers to help build shallow-water galleys to pursue these privateers.[42]

In fact, Bull's desire to expand the colony's small-boat service extended back to July 1740. Immediately following the retreat from St. Augustine in July, the governor informed the Commons House that: "When I consider the Situation of our Southern Frontier . . . by the Spaniards in their Row Galleys, which are capable of coming into any of our Inlets . . . where our larger Vessels cannot get at them; and be ready to intersept any of our Craft, and also encourage the Desertion of our Slaves . . . The best and cheapest Way to disappoint such Attempts would be to [have] 4 or 6 Boats fitted with a 6 Pounder, and several Swivel Guns, Oars."[43]

In essence, Bull not only feared that Spanish light craft would raid local commerce but that they would foment social disarray by encouraging enslaved people to run away from plantations. This warning would have provoked anxiety among the wealthy rice planters that dominated the Commons House of Assembly. After all, only a year before, enslaved Africans along the Stono River—just a few miles south of Charles Town—rose up against planters and killed several inhabitants in what would become known as the Stono Rebellion. The rebels tried to flee south to Spanish Florida, where the governor had promised freedom to those enslaved people that successfully escaped Anglo-American plantations. In their journey southward, the rebels tried to hold off South Carolina militiamen who had been sent to pursue them, but they were overpowered; large numbers of the surviving rebels were put to death.[44]

By mid-December, the Commons House of Assembly agreed to ask the king for six galleys while also reluctantly agreeing to fund local construction of two of the craft. The reasoning for this move is unclear, but perhaps provincial authorities hoped for greater imperial assistance while providing for the possibility of being refused.[45] Whatever the assembly's reasoning, the move to finance local galleys was ultimately sound. While the Board of Trade ultimately forwarded the colony's request for galleys to the Duke of Newcastle, there is no indication he ever agreed to the proposal, and Bull was still campaigning for imperial support by the end of 1741.[46] In the spring of 1741, General Oglethorpe also requested galleys (among other supplies) from the home government and lamented the lack of Royal Navy protection on the Georgia station. Unsurprisingly, imperial officials ultimately did not grant his requests either.[47]

All in all, the British government's opinion on the colonies' provincial navies by the end of 1742 is best summed up by the Privy Council's decision in November of that year to not provide guns for South Carolina's two newly built galleys (*Beaufort* and *Charles Town*). Upon reviewing a plea from Bull to provide nine-pound guns for the vessels, a committee from the Privy Council deliberated on the matter for nearly a year with the colony's agent and with the Ordnance Board and finally came to the conclusion that "as these [row galleys] are intended to Secure the Inland passages of the Province the said Board conceives that the Inhabitants ought to furnish themselves with such Ordnance otherwise the rest of His Majestys Colonys may hereafter Solicit the like favour."[48] This default assumption that Anglo-Americans should fund their own provincial navies mirrored the Admiralty's 1724 aforementioned declaration that "when vessels have been fitted out by the Governors of his Majesty's Islands or Plantations abroad, the inhabitants have borne the charge thereof."[49]

Whereas British authorities expected South Carolina's "Inhabitants" to fund and crew the galleys, it is worth examining who the "inhabitants" of the colony were that crewed these two small warships. The surprising amount of demographic information on the galleys' crews allows us a rare opportunity to examine the diverse backgrounds of sailors within the South Carolina provincial navy. Early on in their service in the summer 1742, neither galley was well manned, and one of the crews on the provincial establishment was as small as three men. Realizing he needed a light craft to accompany him on one of his cruises, Captain Hamar of HMS *Flamborough* felt "Obliged to Man her Out of His Majesty's Ship under my Comand."[50]

By July 1742, provincial authorities hastily impressed and recruited soldiers, sailors, and ships for a relief mission to aid General Oglethorpe as Spanish forces invaded Georgia. Exclusive of Royal Navy sailors, provincial authorities estimated that there were over 560 South Carolina men sent on provincial vessels. Of this number, seventy-three, or 13 percent of the total crews, were listed as "Negroes."[51] Despite the colony's recent experience with a major slave rebellion and restrictions on the rights of free and enslaved Africans to carry weapons, it is clear that South Carolina's ruling class still depended on the enslaved and free African population in the colony to support the war effort against the Spanish.[52] In addition to racial diversity, there is evidence that at least one Jewish sailor (known only as Mr. Hart) served on board *Charles Town* when it sank with its ten man crew in a squall in 1743. Interestingly, Hart was the only man named in the *South Carolina Gazette*'s notice of the tragedy.[53]

While imperial authorities were not quite ready to directly support provincial naval efforts, mercantile anger over Royal Navy negligence was beginning to make an impact at Westminster, and provincial naval activity played a role. Particularly useful for the colony's pleas for naval assistance was a rising lobby of London merchants intimately connected with Carolina's trade and who had powerful connections within the British government.[54] In early 1742, Parliament listened to various testimonies by merchants and ship captains directly affected by alleged Royal Navy negligence in the southern colonies. Agents for and merchants from Virginia and South Carolina claimed that various Royal Navy station captains (particularly Sir Yelverton Peyton and Captain Charles Fanshawe) had been neglectful of their defense, had frequently impressed sailors, had extorted local merchants, and had rarely left port to cruise after the enemy. This inaction, according to the representatives of both colonies, had forced the colonies to fit out their own vessels at great cost to secure their coasts. Notably, one of these provincial guardians—New England privateer Captain John Rous—will appear later in this chapter.[55]

Ultimately, the parliamentary committee that heard the complaints concluded that "due and necessary care had not been taken to keep a proper number of his Majesty's ships employed" in protecting English commerce and wanted the rest of the House of Commons to "bring in a bill for the better protecting and securing the trade and navigation of this kingdom."[56] Historian H. W. Richmond has argued that Admiralty resistance led to the bill's failure, as it "contained stringent clauses to tie the stationed ships securely to their stations and allow their Captains no liberty of action." In practice, the bill would have given colonial governors and councils near total control over station ships' orders. Despite the legal failure, with rising complaints from many different colonies over Royal Navy performance, stronger instructions were given to Royal Navy captains to work with governors.[57]

With significant pressure from trading interest groups and increasing numbers of ships lost to Spanish privateers in the Americas, imperial authorities committed to more than just a change of rhetoric in instructions to new Royal Navy station captains. The 1742 edition of the *Scots Magazine* reported that an Admiralty court martial commenced on June 9, 1742, and "adjudged Sir Yelverton to be dismissed as a Captain of the Royal navy; and adjudged Captain Fanshaw to be mulcted six months pay for the use of the chest at Chatham [Naval Hospital]."[58] Even before court martialing these captains, Admiralty officials had dispatched Captain Charles Hardy with HMS *Rye* (alongside the sloop *Hawke*) to replace Fanshawe and gave him specific instructions to be more proactive in hunting Spanish privateers and convoying merchant vessels than his predecessor.[59] While provincial authorities would never be completely happy with Royal Navy station captains, these actions did signal that the British government was beginning to take coastal defense in its North American colonies more seriously.[60]

Meanwhile, General Oglethorpe in Georgia would have appreciated royal assistance of any kind. Not long after the defeat at St. Augustine, Oglethorpe formed a "Marine Company of Boatmen" with recruits from Maryland, Virginia, and South Carolina. Whatever the novelty of American marines serving in an Anglo-American navy, the general's small craft alone could not take on Spanish privateers at sea. With most British warships staying centered at Charles Town, Oglethorpe had expanded his own provincial navy to include the schooner *Walker* along with the *Faulcon* and *St. Philip* sloops. He had purchased these ships in Charles Town, crewed them with South Carolina sailors, and used redcoats from the British Army's Forty-Second Regiment of Foot to act as marines on these larger vessels.[61]

Indeed, Oglethorpe had every reason to be confident in these provincial forces. In a December 1741 letter to the Georgia Trustees' accountant, Oglethorpe had justified his provincial navy's existence by detailing the inability of Royal Navy station ships to stop enemy privateers throughout the region. He then bragged that his own provincial forces had "already forced one of the Enemys Sloops on Shore."[62] In a June 1742 "List of the Military Strength of Carolina & Georgia," Oglethorpe reported thirteen vessels in the colony's service, ranging from the "guard schooner" to small boats that various infantry regiments used. Excluding soldiers that manned the latter craft, Oglethorpe's provincial navy exceeded one hundred sailors. It is telling that he excluded South Carolina's scout boats or galleys and briefly mentioned the "Men of War Stationed at Charles Town" at the end of the report.[63]

Around the same time Oglethorpe filed that report in June 1742, the general received intelligence of a large Spanish invasion force that was likely headed for coastal Georgia, and he forwarded the news to Charles Town. Oglethorpe's successful defense of the colony highlighted further flaws in the uneasy relationship between provincial and Royal Navy forces. The invasion threat prompted Lieutenant Governor Bull and his council to summon Captain Hardy and Captain Franklin of the recently arrived HMS *Rose* on June 18 to discuss the best method to assist their southern neighbors. Though Hardy informed the council that his ship was too damaged to sail south at that time, Frankland offered to take his vessel, HMS *Flamborough* (now captained by Joseph Hamar), and the *Charles Town* galley with him on the way back to his own home station in the Bahamas.

Rather than keeping his promise, Frankland abandoned the flotilla early on, and Hamar himself took his own vessel, two small Royal Navy sloops, and *Charles Town* to assist Oglethorpe. Hamar made it to St. Simons, Georgia, by July 13 but immediately ordered his flotilla to retreat back to South Carolina when he sighted the numerically superior Spanish invasion force of forty vessels. Captain Hardy—having finished *Rye*'s repairs—led a subsequent joint provincial-royal relief force with six colonial vessels, only to discover that the Spaniards had fled by that point. Much to the chagrin of the South Carolina government, Hardy ordered the provincial vessels to return home and decided against hunting for what remained of the Spanish forces as he feared they may have sailed to attack South Carolina's weakly defended Port Royal district.[64]

While the provincial relief force that accompanied Hardy did not see much action, the fact that South Carolina's legislature was able to send out six armed vessels with more than six hundred sailors testifies to the colony's growing provincial naval establishment. Each provincial naval captain was given a letter that ordered them to obey orders from Captain Hardy. Bull and his council also commanded each captain to follow "Articles and Orders for the regulating and better [Government] of the Vessels & Forces by sea fitted out from Charles Town . . . pursuant to the direction of the Statute of the 13th of Charles the 2d: Chapter 9th."[65] With that seventeenth-century statute for Royal Navy ships as a guide, they would have included provisions such as ensuring public worship, punishments for misbehavior, rules for prize distribution, and so on. In essence, the South Carolina government considered its provincial naval forces subordinate to Royal Navy authority but also bound by the same standards and rules the Royal Navy operated under.[66]

While South Carolina's provincial forces and their Royal Navy allies sailed confusedly back and forth between both colonies, Oglethorpe's small army of one thousand men and even smaller provincial navy fended off a Spanish army twice its size, as well as a substantial Spanish fleet. After scattered skirmishes with enemy forces for several days, Oglethorpe found enough time to organize his infantry forces at strategic locations. By July 5, Spanish governor Montiano led his force of thirty-six vessels ranging from ships to galleys into St. Simons Sound and set the scene for one of the largest battles an Anglo-American provincial navy would ever engage in.

Aside from a few of his own privately owned vessels, Oglethorpe had impressed several merchant ships and their crews in the sound and fitted out the largest—*Success*—as his flagship. Oglethorpe added twenty guns on board and crewed it with sailors, British regulars, and his own provincial marine company.[67] Though his naval forces were not adequate to stop the Spanish onslaught, Oglethorpe's land forces defeated the Spanish invaders in several engagements throughout the next several weeks and forced them back to Florida. For now, the British hold on Georgia was secure.[68]

While Oglethorpe's victory over the Spanish invasion was noteworthy, South Carolina authorities criticized Royal Navy captain Hardy's "returning hither, before so considerable a part of the Service as the destroying the Enemys Strength by Sea: And for which Our Shipping were fitted out at so considerable an Expence."[69] Some scholars have attributed the colony's dispute with Hardy to the well-established acrimony between the colony and Royal Navy officers, as well as disputes over who had authority over station ships.[70] While it is true that tensions were rising, it is important to note that provincial authorities did still show deference to Royal Navy commanders. Afterall, a contemporary Pennsylvania newspaper described a much smaller raid on Spanish Florida that summer in which "Provincial Vessels [from South Carolina] received their orders from the Commodore."[71]

Even this level of moderate cooperation between the Royal Navy and provincial forces would do little to assuage Anglo-American authorities, who were growing irate over the growing costs of naval defense as the war dragged on. By mid-1743, Oglethorpe had gotten little assurance from London that his extensive military expenditures would be reimbursed, so he traveled to England to directly appeal his case.[72] In March 1744, Oglethorpe—himself a veteran member of Parliament—spoke before the House of Commons, highlighted the tenuous position of the empire's southernmost American colony, and effectively convinced his fellow legislators to reimburse his expenditures of more than £66,000.[73] More than £22,000 of the reimbursement directly covered maritime purchases, upkeep, and pay for boatmen and sailors.[74]

While Parliament reimbursed Oglethorpe, it also officially placed Georgia's soldiers as well as provincial naval forces on a royal pay establishment similar to that of the British army.[75] For the first time in colonial history, the imperial government officially supported a provincial navy; the same royal government that built first-rate warships at Portsmouth, England also bankrolled Anglo-American schooners and scout boat crews in coastal Georgia.

Limited Support: The Growth and Limits of Imperial Support for Provincial Navies, 1744–1754

Although the war with Spain raged on, continental tensions drove Britain and France into open conflict in 1744 in what has become known as the War of the Austrian Succession (or "King George's War" in America). The British Empire now faced an even larger global and multi-theatre conflict.[76] That same fateful year, parliamentary financial support for provincial naval forces would also extend to Nova Scotia and New England, and the Admiralty would finally encourage more active naval campaigns in North America.

On the surface, 1744 appears to be a pivotal moment when imperial officials—embattled by an ever-growing web of external foes—recognized their provincial constituents' needs and acted to remedy their previous inattentiveness to colonial naval defense. Nevertheless, inconsistent imperial policies regarding the legal status of provincial navies—combined with increasingly aggressive and undiplomatic behavior by Royal Navy station captains—damaged this burgeoning royal-provincial naval partnership.

Legal battles related to provincial navies, prize money, and impressment were the results of various parliamentary acts such as the 1740 "Act for the more effectual securing and encouraging the trade of his Majesty's British subjects to America, and for the encouragement of seamen to enter into his Majesty's service" (13 Geo 3, c. 4), the 1744 "Act for the Better Encouragement of Seamen in his Majesty's Service, and Privateers, to Annoy the Enemy" (17 Geo. 2. c. 34), and the 1746 "Act for the Better Encouragement of the Trade of His Majesty's Sugar Colonies in America" (19 Geo. 2, c. 30). While some provisions of these acts increased the total amount of prize money available for royal and privateer sailors who captured enemy vessels, they also expanded the government's ability to impress mariners (outside of the Caribbean) and attempted to expand Westminster's oversight over the privateering enterprise in general.[77]

Confusingly, even as Parliament expanded its support for provincial navies in Georgia and elsewhere, it never included those forces in these acts. Vague language such as "encouragement of the officers and seamen of his Majesty's ships of war, and the officers and seamen of all other British ships and vessels, having commissions, or letters of marque" did not directly recognize colonial governments' own naval forces even though Parliament was well aware of them by now.[78]

This is not to say that provincial governments were always consistent in this matter; there were of course still cases where Anglo-Americans themselves still failed to differentiate between commerce raiders with letters of marque and government-funded warships, calling both "privateers" at random.[79] Nevertheless, provincial governments were increasingly using different terminology for their state-run fleets such as "country vessel" or "guard sloop." Furthermore, Westminster's failure to include the colonies' regular naval forces in prize court legislation would lead to legal battles between agents for New England's provincial navies and the Royal Navy for years to come.

These legal and operational tensions would come to a head as the momentum of the war shifted to the empire's northernmost colonies near French Canada in 1744. While northern legislatures had outfitted navies to fight Spanish privateers in the first few years of the war, the opening of hostilities with France in 1744 sent the region into a panic due to the proximity of French Canada. With earlier notification of the commencement of hostilities with their English foes, the French governor of Louisbourg on Isle Royale, the Seigneur Duquesnel, dispatched two privateers and an invasion force to attack the Anglo-American base at Canso. The force quickly captured the settlement, as well as its solitary Royal Navy guard sloop.[80]

Almost immediately, authorities throughout the northeastern colonies mobilized their provincial naval forces to counter the French onslaught. Massachusetts governor William Shirley's quick dispatch of the provincial snow *Prince of Orange* with soldiers played a signal role in repelling Duquesnel's forces from taking Annapolis Royal that summer.[81] Massachusetts's quick response was a result of the maturation of what could be called New England's provincial naval network. After nearly a century of commissioning tax-funded ships, Massachusetts, Rhode Island, and Connecticut each could boast of a complex "naval establishment" of sorts that involved bureaucratic government committees that procured vessels and regulated pay for officers and sailors.[82]

This provincial naval network had reached such a stage that by mid-1744, Rhode Island's government was able to facilitate several joint patrol cruises between its province sloop, *Tartar*, and Connecticut's province sloop, *Defence*. At one point, Captain John Prentice of *Defence* even made the friendly boast that "We can out sail the Rhode Island sloop much. . . . We beat their tip top boats at Rhode Island to their great mortification."[83]

Despite this cooperation between New England governments early in the war, the region's provincial navies suffered from the Royal Navy's largest ailment: perennial manpower shortages.[84] To offset this issue, in the summer of 1744, the Massachusetts governor, council, and assembly passed the "Act for the more effectual guarding and securing our Sea Coasts, and for the Encouragement of Seamen to enlist themselves in the Province Snow or such Vessels of War as shall be commissioned and fitted out by this or other of his Majesty's Governments." This act granted sailors of provincial warships total claims over captured French shipping and cargo, three-pound bounties for the capturing or killing of enemy sailors. The act also awarded the same bounties to provincial navy crews from other colonies, privateers, and merchant ships with letters of marque for every enemy sailor captured or killed off the Massachusetts coast.[85]

What was extraordinary about this law was not only that it was the first time a colony promised financial rewards to other colonies' provincial navies but also that it essentially copied elements of the British government's 1740 "Act for the better Supply of Mariners and Seamen to Serve in His Majesty's Ships of War." The act promised five pound prizes "unto the Officers, Seamen, Marines . . . Onboard such of His Majesty's Ships of War, as also of Privateers" that followed the aforementioned 1740 privateering act.[86] While the British act mentioned Royal Navy ships and privateers, the Massachusetts law specifically targeted the crews of provincial "Vessels of War." For the first time, authorities from one colony offered to support the provincial navy of another.

While Massachusetts may have adapted imperial standards for its own provincial naval establishment, the similarities with parliamentary legislation caught the attention of the Board of Trade. What ensued was the first of many inconsistent imperial rulings on the status of provincial navies. In October 1744, the Board of Trade requested Francis Fane—a member of Parliament and a commissioner for the board—to compare Massachusetts's law with the British government's various bounty laws, and to decide if *Prince of Orange* and similar vessels "are to be deemed ships of war or privateers, and whether they are entitled to the bounties given by the said British acts."[87] Fane did not oppose the act by "point of law" but worried over the vagueness of the act. He argued that the "Province Snow and the other Vessels mentioned in the said Massachusetts Act, will be Entitled to the Bounty given by the said British acts . . . as Privateers because they are not in his Majesty's Pay." Fane also complained that Massachusetts had "gone a little too far in disposing of His Majesty's right to the Prizes taken from the Enemy, solely by their own Authority."[88] Ultimately, Fane argued that provincial ships were privateers if they were not on the royal payroll— perhaps an oblique reference to the government's recent funding of the Georgia navy.

While Fane dismissed Massachusetts's provincial navy as a fleet of privateers, the Board of Trade was still uncomfortable with simply dismissing the law and decided to table the debate until Governor Shirley and Massachusetts's agent in London could better explain it. By the spring of

1747, the King's Privy Council reviewed the act that had been in bureaucratic limbo for two years. The Privy Council concluded that the act "relates to the public service & security of the said Province and therefore We see no reason why His Majesty may not be graciously pleased to confirm the same." The king ultimately agreed with the Privy Council and confirmed the act by June 1747.[89]

Even though the royal approbation of the Massachusetts law took several years, it highlighted two conjoined trends in the latter years of the fight against Spain and France: London's increasing recognition of provincial navies, and its confusion over how to classify them or incorporate them in the larger war effort. At the same time, Admiralty officials continued their age-old "laissez-faire" attitude toward provincial navies, particularly when a "hands-off" approach to colonial naval defense could save them money. For instance, in the spring of 1745, Thomas Corbett—the secretary to the Admiralty—informed Commodore Peter Warren that their "lordships hope that your letter to Gov. Clinton, about the province of New York supporting a guard vessel as the other neighbouring colonies did, has had its due effect."[90]

While Admiralty officers hoped that provincial authorities would continue to fund their own navies, some Royal Navy officers—particularly Warren—began to lobby for greater imperial support for provincial naval forces. He called for colonists to begin: "arming some proper vessels to guard their own coast and trade. [Such] vessels should be in some measure on the foot[ing] of the king's ships, or at least [should] never be molested by them.... Where the colonies are not in a capacity alone to bear the expense of such vessel, two or more of them might join in it.... This I believe, the colonies [might] be brought to, if strongly recommended by the ministry to their different governors, and by them to their legislatures. New England has shown the others a very laudable example, by fitting out two or three. If this could be effected, then his Majesty's ships of force ... might be employed in distressing the enemy more effectually."[91]

Extraordinarily, Warren not only encouraged the British government to categorize provincial fleets "on the same footing" as royal frigates but also supported the idea of multicolony fleets—a suggestion that would not be taken seriously by American authorities until 1775. Warren's unique support for provincial naval forces would fully crystallize in the joint provincial–Royal Navy siege of Fortress Louisbourg in 1745. This siege, holds historian W. A. B. Douglas, "demonstrated the surprising strength and homogeneity of combined regular and provincial forces."[92] While the siege itself would demonstrate the potentials of such cooperation, widespread interservice tensions in its aftermath challenged any future cooperation.

After the British government refused to spearhead an attack on Louisbourg in 1744, Governor Shirley (with the lobbying of Maine merchant William Vaughan) convinced the Massachusetts legislature to lead an assault on the French stronghold in the spring of 1745. Without direct guarantees of British military assistance, he hoped that neighboring colonies would join the expedition and that the British government would reimburse the colonies for taking France's Canadian privateering base. Initial Massachusetts plans called for a joint expedition of provincial land and naval forces from every northern colony, but this strategy was overly optimistic. Excepting a single artillery battery from New York, most every provincial soldier and sailor at the siege came from New England.

By May, there were around one hundred merchant vessels from New Hampshire, Massachusetts, Connecticut, and Rhode Island ready to transport the region's infantry forces.[93] Included in this flotilla was a squadron of fifteen provincial naval vessels and privateers with a combined strength exceeding one thousand sailors.[94] Undoubtedly, one of the most impressive provincial vessels was Massachusetts's recently constructed four-hundred-ton, twenty-gun frigate *Massachusetts*.[95]

Even with the largest provincial naval force assembled to date, Shirley did not believe the expedition would be successful without Royal Navy assistance. In a late March letter to the Duke of Newcastle, Shirley described the New England colonies' vast military preparations and complained that Royal Navy officers in the West Indies were not able to assist the expedition. He continued: "I shall hope that Providence will favour the small Naval Force, which I have been able to muster up here, with Success; and that our Land Forces will still be able to maintain their ground on Cape Breton 'till I shall receive his Majesty's Royal Pleasure upon this matter."[96]

Luckily for Shirley, changes in the Admiralty's administration (including the Duke of Bedford's appointment as the First Lord of the Admiralty, as well as Admiral George Anson as one of its commissioners) may have played a role in policy changes in North America. In early 1745, the Duke of Newcastle ordered the creation of the first ever North American Squadron—a move that united British vessels north of the Carolinas under one command. By March, the Lords of the Admiralty received word of Shirley's preparations and ordered Commodore Warren to assist the provincial forces in taking Louisbourg. When Warren arrived with ten Royal Navy warships and instructions from Newcastle to take command of provincial "shipping," Governor Shirley and the provincial military leader William Pepperell placed the Anglo-American vessels under Warren's command—the same decision made by South Carolina authorities during the campaigns of 1740–42.[97]

In many ways, Warren (himself a veteran of the disastrous St. Augustine campaign) did his best to ensure cooperation between provincial and Royal Navy forces. W. A. B. Douglas contends that "there is strong evidence that both Warren and Shirley intended to consider the armed colony cruisers and king's ships as a homogenous squadron attached to the expedition." To support this claim, Douglas highlights Warren's inclusion of provincial ships in his line of battle, his inclusion of provincial commanders in councils of war, and the fact that Warren ordered Royal Navy and provincial navy crews to distribute captured prizes equally—a conciliatory tactic never tried by other Royal Navy commanders.[98]

After a month of deadly assaults, bombardments, and raids, the French garrison at Louisbourg surrendered to the joint invasion force on June 17, 1745. Governor Shirley bragged to the Massachusetts legislature that Louisbourg "was won, under the most signal Favour and Direction of the Divine Providence, by the indefatigable Toil of His Majesty's New-England Subjects (chiefly of this Province) supported by a Squadron of his Ships of War at Sea."[99] Perhaps one of the most concrete examples of the fruits of this partnership was New England privateer Captain John Rous's promotion. Rous was a New England privateer who had previously acted as a coast guard for South Carolina authorities and now served in the provincial navy of the New England invasion force at Louisbourg. Admiralty authorities were so impressed by news of his fight with a French frigate during the siege that they commissioned him as a captain in the Royal Navy, purchased his vessel—*Shirley*—and made it an official part of the navy.[100]

For a moment, British strategic planning seemed keen on including provincial naval forces. This is evident in First Lord of the Admiralty, the Duke of Bedford's plans for an abortive 1746 conquest of French Canada. Bedford ordered that royal forces should be accompanied by "such ships of war, sloops and such other armed vessels (which may be furnished by the provinces) as his Majesty's admiral commanding in chief shall please to appoint."[101] Even when this joint expedition was canceled by imperial authorities, Parliament still reimbursed the New England colonies for their military preparations (including naval expenditures) for the siege of Louisbourg and the canceled 1746 expedition.[102]

The evidence presented thus far may create the appearance of a growing and unreserved spirit of support and approbation of provincial navies in the imperial center after 1744. However, this growing imperial enthusiasm for provincial forces was inconsistent and often shallow.

Chief among the flaws in the arrangement was the fact that the British government never created a permanent legal standard or definition for these provincial fleets. Legal uncertainties over the status of provincial fleets fostered bitter transatlantic legal battles, particularly after the victory at Louisbourg. A few weeks after the city fell, provincial naval forces captured several French prizes both independently and alongside the Royal Navy. With Commodore Warren's promise that the joint fleet would share in the "common stock" of any prizes captured, questions immediately arose over whether seamen in Royal Navy ships and provincial vessels should have an equal share of the booty. Beyond mere disputes over plunder, major controversy arose over the very nature of provincial navies themselves and whether provincial vessels should receive the same prize shares as Royal Navy ships or privateers.[103]

The next salvo in the transatlantic dispute over the definitions of provincial navies came in the Massachusetts Vice Admiralty court of Robert Auchmuty in the early months of 1746. Auchmuty, like the vice admiralty judges of other colonies, was not merely a provincial justice but "officially appointed at Whitehall with Admiralty warrants."[104] Although Auchmuty was a veteran jurist with training at the Middle Temple in London he was still ill prepared for the prize claim of Captain Richardson and the crew of *Resolution*—a private sloop leased to the Massachusetts government for the expedition against Louisbourg.[105]

On September 2, 1745, Richardson and his crew recaptured an English vessel called *The Two Friends* from the French. Richardson and his men declared themselves the crew of "his Majesty's Vessel of War and in his Majesty's Pay" and thus entitled to "one Entire Eighth" of the vessel's value as a salvage fee.[106] They based their claim on Parliament's 1744 "Act for the Better Encouragement of Seamen in his Majesty's Service," which guaranteed Royal Navy vessels one-eighth of the value of a recaptured English vessel no matter how long it had been in enemy hands. Conversely, the act merely granted privateers that recaptured English vessels shares ("moieties" of the value) that decreased by percentage the longer the English vessel had been controlled by the foe.[107] It was clear to Richardson and his men that it would be more profitable to be counted as part of the king's navy rather than as mere privateers.

Unfortunately for these provincial sailors, Judge Auchmuty was not convinced by their claim and held that "Every Kings Ship is in his pay and Service and part of his Royall Navy but Every Ship in the Kings pay and Service is not the Kings Ship or part of the Royal Navy." Auchmuty examined the history of private ships in the royal service as far back as Edward III's reign, more recent parliamentary legislation, and then Captain Warren's specific prize agreement for his joint fleet, and found nothing to support the argument that *Resolution* was a royal ship of war. Auchmuty reasoned that Warren's instructions could not be construed to equate *Resolution* with Royal Navy ships in the fleet because he "treats those Vessels in Contradistinction to his Majestys Ships by sometimes Calling them private Ships & Vessels of War and at other Times Colony Cruizers." Furthermore, Auchmuty reasoned, *Resolution*'s owners still expected a share despite contracting her to the government—a move that highlighted the ship's status as a privateer rather than a vessel of war. The Vice Admiralty court ruled that *Resolution*'s men could only expect the privateer share of their prize.[108]

At least some provincial elites were taken aback by the ruling. Nathaniel Sparhawk, the son-in-law of William Pepperell (commander of provincial infantry forces at Louisbourg and one of the presumed investors behind *Resolution*), wrote his father-in-law to lament that "She is, contrary to the expectation of every one, deemed a privateer instead of a King's ship" and worried that appealing the case in London would cost Pepperell more money than it was worth.[109]

While Auchmuty's ruling was unpopular with some of the expedition's provincial leaders, this was merely the beginning of what would become a series of transatlantic disputes over prize money. Those larger fights involved a series of prize-court disputes following the joint Royal Navy–provincial capture of the French supply ship *Notre Dame de Deliverance*. This ship, along with its sister ships *Heron* and *Charmante*, fell prey to Anglo-American forces at Louisbourg in the weeks after the city's capture. In August 1745, Captain Benjamin Fletcher of the provincial brigantine *Boston Packet* spotted what he thought was a French frigate. Realizing his small crew could not take on such a vessel alone, he raised a French flag as a decoy, fired guns to alert the Royal Navy ships nearby, and fled for the protection of Louisbourg. The Royal Navy frigates *Chester* and *Sunderland* quickly captured the "frigate," which turned out to be a treasure-laden vessel worth nearly £400,000.[110]

For more than four years after this lucrative capture, agents for *Boston Packet*, *Chester*, *Sunderland* and other nearby provincial and Royal Navy ships tried to convince Admiralty appeals courts in London of their competing claims to the prize.[111] While interested parties argued over which vessels were most responsible for the capture of *Notre Dame de Deliverance*, questions over the legal status of the provincial vessels arose time after time.

In fact, representatives for the provincial vessels argued that their clients belonged to vessels that were essentially ships of the Royal Navy. For instance, in one of the earlier hearings in May 1749, Charles Pinfold, one of the advocates for *Boston Packet*, argued that the vessel was no privateer as the colony's government had purchased her. Pinfold continued that "Privateers are fitted out at Private Expence with Letters of Marque, Security is given, and an Agreement made with the owners." Additionally, Pinfold made the accurate observation that no recent parliamentary legislation differentiated between a "Man of War and a Vessel in his [Majesty's] Pay."[112]

In a subsequent hearing, agents for some of the other provincial vessels echoed Pinfold's argument when they said their clients were "not Privateers, belonging to particular Owners, but were Ships of War, of considerable Force, fitted out by the Colonies of the Massachusetts Bay and Rhode Island." As ships of war, they had played a signal role in the siege and as part of Warren's fleet.[113] While these arguments were made by legal counsels in London, they clearly represented the belief among many provincial authorities and sailors that provincial navies should be considered part of the king's standing military forces rather than as privateers.

To the contrary, agents for the Royal Navy frigates involved in *Notre Dame de Deliverance*'s capture had a different understanding of the role of provincial vessels in the expedition. In one hearing, a Royal Navy counsel argued that "American Privateers, by their Junction with Sir Peter Warren, became no otherwise Part of his Squadron, or subject to his Command." While spending much time decrying the provincial men as mere privateers, the agents for the royal frigates did make one thorough accusation against their opponents: If the point of the expedition was to capture Louisbourg, and the capture of *Notre Dame de Deliverance* occurred after the fact, why should the provincial ships be considered part of a joint squadron? After all, the provincial governments had already been "reimbursed by the Parliament of Great Britain" for fitting out warships to take Louisbourg.[114]

Ultimately, after several years of lengthy litigation, on July 5, 1750, the Lords Commissioners of Appeal for Prizes—including members of the Admiralty's new cohort of reformers, Lords Anson and Sandwich—ruled that the Royal Navy vessels in Louisbourg harbor "in sight" of *Notre Dame de Deliverance*, as well as the two Royal Navy frigates that captured the French vessel were all entitled to shares of the prize. The Lords specifically excluded the other American "privateers" that made claims on it. Nevertheless, by November the lords did declare that the "armed vessel" *Boston Packet* and the Royal Navy warships should all receive equal shares.[115] The Admiralty had awarded an American provincial crew an equal share to the royal crews but deftly avoided calling the American vessels "Ships of War."

In recent years, some scholars have highlighted the wedge this case would create between the British government and provincial military forces. For instance, naval historian J. Revell Carr has highlighted the connections between this case and an anonymous 1748 essay (possibly written by the Boston firebrand Samuel Adams). The anonymous colonial author lambasted the British government for inadequate naval patrols off the New England coast and for not sharing plunder from the three captured French vessels with New England infantrymen and critiqued British soldiers for abusing Anglo-Americans during the siege. Britain's most damning affront, the correspondent argued, was its recent decision to return Fortress Louisbourg to the French in the peace negotiations at Aix-la-Chapelle. Carr contends that the letter was emblematic of the growing "seeds of discontent that [would bring] the Americans to the brink of revolution."[116]

Whether or not the *Notre Dame de Deliverance* case mattered to future revolutionaries, it did have immediate implications for the fraught military partnership between British forces and their provincial counterparts. The years of legal battles over the prize's fate demonstrated two diverging imperial and provincial views of the importance of provincial navies. While imperial officials were finally starting to encourage the colonies to build provincial fleets (and even occasionally funding them), they still only thought of these vessels as auxiliaries for the Royal Navy. Thus, they never even made room for provincial navies in imperial legislation.[117]

On the other hand, many Anglo-Americans were increasingly coming to see their provincial navies as equals to the Royal Navy. This mood is best illustrated in a late 1747 letter from the young South Carolina merchant Henry Laurens to his colony's agent in London. Laurens, who would one day be a founding father of the United States, bragged that "we are fitting out two fine Bermuda Sloops on purpose to Cruize on this Coast. . . . As to [British] Men of War, they are out of fashion here."[118] It is noteworthy that Laurens, the scion of an aristocratic South Carolina family, envisaged Royal Navy ships as being "out of fashion" like a passing gentleman's clothing trend from London!

These sentiments were not only the purview of North American colonists but were also shared by some of their West Indian compatriots. For instance, in the summer of 1746, England's *Gentleman's Magazine* published a letter from an Antiguan who lamented that because of the "indolence of his majesty's ships . . . the country have fitted out a guard de costa." The correspondent, echoing contemporary complaints on the North American mainland, further alleged that they "pretend that they cannot sail well enough to catch the privateers; but all the world knows, that they can sail well enough to protect and retake the merchant ships, if they would keep cruizing in proper stations."[119]

While Caribbean governments still occasionally fit out provincial vessels to augment royal forces, it is also true that the War of Jenkins' Ear saw the maturation of the British government's ever-increasing investment in the naval infrastructure of its Caribbean provinces. Historian Richard Harding argues that this conflict was the "first time it had proved possible to maintain reasonably large squadrons" in the Caribbean, with victualing and careening bases at Port Royal, Jamaica and the two-decade-old base at English Harbour, Antigua. While the Royal Navy still depended on mainland North American sailors and supplies to sustain their West Indian fleets, these facilities gave Royal Navy squadrons in the Caribbean greater flexibility than their Franco-Spanish foes.[120]

That the Royal Navy was more engaged in the West Indies than in North America can be demonstrated by examining the disparity of warships between both regions. For instance, in late 1742, while there were only ten Royal Navy guard ships and smaller vessels stationed throughout the entirety of North America, there were twenty-five war vessels in the Jamaica squadron alone, and eight others stationed in the Leeward Islands and Barbados.[121]

Even if one could make the case that it was necessary to have a larger Royal Navy presence in the Caribbean because of the numerous Franco-Spanish enemies surrounding the British islands on all sides, parliamentary legislation during the war clearly demonstrated London's preference for its West Indian colonies. Its aforementioned Act for the Better Encouragement of the Trade of His Majesty's Sugar Colonies in America disallowed Royal Navy impressment in the Caribbean but sanctioned it in the mainland North American colonies. Historian Christopher Magra writes that this "new impressment legislation created stake-holders in the Caribbean. Sugar planters could rest more secure . . . as press gangs would no longer disrupt the flow of slaves from Africa and provisions from around the Atlantic world."[122]

This short-sighted and regionally discriminatory law would lead to a wave of violent resistance in war-weary northern ports. Of course, as we have seen, violent tensions over impressment were hardly a new phenomenon in the Atlantic world in 1746. Nevertheless, this ill-timed act came at a time of amplified resistance to the practice in New England In November 1745 when Lieutenant Governor Spencer Phips allowed Captain Arthur Forest of HMS *Wager* to impress a few men, provided they were nonresident aliens and had not served in the Louisbourg expedition. Ignoring this prohibition, the ship's press gang (along with local sheriffs) attempted to capture several sailors who had served on the provincial vessel *Resolution*. A subsequent melee left two provincial sailors dead and three members of the press gang in provincial custody; the rest of the press gang escaped with *Wager* as it left Boston harbor. Historian Jack Tager contends that this violent encounter would be a "rallying cry" for the rioters in the better known Knowles Riot two years later.

While Governor Shirley vocally opposed the violence, local officials in Boston criticized him and his administration for allowing impressment in the first place and called it a violation of the Magna Carta and parliamentary legislation. Although Shirley convinced Commodore Warren to cancel further calls for impressment throughout the Northeast, Royal Navy commanders ignored this directive and continued to rely on the unpopular practice to keep their ships fully crewed. Even locals were not safe from resistance to impressment. When newly minted Royal Navy captain John Rous tried to impress sailors for HMS *Shirley* in February 1746, angry locals (along with a privateer crew from New York) assaulted Rous and his press gang.[123]

None of these violent clashes in the final years of the War of Jenkins' Ear could compare to the Knowles Riot of 1747. When Commodore Charles Knowles prepared to sail to the West Indies in late 1747, he stopped to impress sailors in Boston because of the recent parliamentary legislation that had banned impressment in the Caribbean. Nearly three hundred angry privateers violently resisted those press gangs—an act that would ultimately inspire a general urban riot that would last for three days. Bostonians imprisoned Royal Navy officers, destroyed one of the royal frigates' barges, and forced Governor Shirley to provide refuge for some Royal Navy commanders in his home.

Soon thereafter, the rioters demanded that Shirley not only deliver them the officers hiding in his house but also that he execute one of the still-imprisoned members of the press gang that had killed men from the provincial warship *Resolution* two years previously. Shirley declared he would wait for the king's instructions before putting anyone on trial and did his best to assuage the crowd by promising to get the recently impressed New

Englanders released.¹²⁴ Shirley later reported that along with other local dignitaries, Captain Edward Tyng of the provincial *Massachusetts* frigate "stood some time at the Door parlying and endeavouring to Pacify 'em."¹²⁵ In a moment of pure historical irony, a prominent provincial navy captain attempted to defend Royal Navy officers from a crowd still angry that Royal Navy mariners had killed provincial sailors.

As the violence of the riot escalated, Governor Shirley only barely restrained Commodore Knowles from ordering his ships to fire on the port of Boston. Ultimately, representatives from several factions convened a town meeting, condemned the mob (much to the dismay of the young firebrand Samuel Adams), and arranged for the release of the impressed sailors. Historian Christopher Magra contends that the Knowles Riot was a precursor of sorts to the violent riots of the Revolutionary era less than two decades later.¹²⁶

While this riot may have served as a template for future American resistance to British authorities, it is also certain that much of the original animus that led to the Knowles Riot surrounded Royal Navy abuse of provincial navy veterans. Even though some British authorities hoped to attract larger provincial naval support for late-war expeditions, parliamentary legislation that had condoned widespread impressment undermined any long-lasting naval partnership with Anglo-American forces.

Even as Anglo-American resistance to Royal Navy impressment increased in Boston in the final years of the War of Jenkins' Ear, imperial authorities began to slowly cut back their support for large provincial navies and developed a laissez-faire attitude toward the few small provincial naval forces they did bankroll during peacetime. Even after General Oglethorpe had returned to England, the British government had continued to pay for Georgia's provincial flotilla—including Oglethorpe's provincial marine corps, a merchant ship converted into a frigate, a schooner, a sloop, a periagua, and sundry boats. With peace overtures already beginning by 1746 and in response to shoddy bookkeeping by provincial officers, the War Office suspended all support for provincial naval forces in Georgia outside the crew of one scout boat, *Prince George*.¹²⁷ The Crown did offer South Carolina authorities three boats in 1749, but the local assembly (much to the chagrin of Governor John Glen) decided to take the expense of fitting out the vessels themselves in order to expedite naval patrols for escaped enslaved people.¹²⁸

While Parliament did agree to purchase some scout boats on the southern borderlands of the continent, it continued to bankroll a more substantial fleet on the Nova Scotia frontier. Provincial vessels had assisted the Royal

Navy off of Nova Scotia for several years after the campaign at Louisbourg in 1748. Nevertheless, by 1748, the Admiralty ordered Governor Shirley of Massachusetts to disband two hired vessels (*Anson* and *Warren*) as it planned to cut costs. Shirley refused to follow the order as he believed they were still necessary for coastal security, and the Board of Trade eventually agreed to help find funds for the vessels.[129] In fact, when George Montagu-Dunk, the Earl of Halifax—the new president of the Board of Trade—planned a new settlement in Nova Scotia after the war, he consulted provincial naval veteran/Royal Navy captain John Rous and even agreed that the new colony needed to employ three provincial guard sloops. Rous would later be appointed the "senior naval officer" of the new port.

Even with Rous at the helm in Halifax, the ever-parsimonious Admiralty refused to keep a sufficient Royal Navy presence in the Canadian Maritimes, and Rous and local political officials relied on a "sea militia" of several small sloops and other vessels to guard the coasts. This was especially important as tensions with the Mi'kmaq flared up in the early 1750s. From 1749 to 1755, Parliament bankrolled eleven provincial schooner, sloop, and boat crews on the empire's northern American borderlands. Throughout this period, Parliament did not assume any serious oversight responsibilities for these forces. Ironically, historian W. A. B. Douglas argues that the Admiralty's noninvolvement was the "essential ingredient of success" in this arrangement.[130] In essence, imperial funds were important for the Nova Scotia sea militia, but London's noninterference allowed local officials to control the fleet to their best advantage.

From 1739 to 1748, the British government agreed to finance some provincial naval forces for the first time, thereby creating the potential for a mutually beneficial naval defense partnership between periphery and center. Nevertheless, the British government's failure to include provincial navies in major sea-prize legislation and its failure to limit Royal Navy impressment damaged cooperation between provincial and metropolitan forces and left a sour taste in the mouths of colonists for the Royal Navy in general. Imperial inconsistencies and Royal Navy overreach would continue to plague Anglo-American relations for the few remaining decades before the American Revolution.

5

The Replacement of Provincial Navies by the Royal Navy in the Seven Years' War, c. 1756–1763

The War of Jenkins' Ear/King George's War ended in 1748 the same way many of the earlier colonial conflicts had: a return to the *status quo ante bellum*. On the tense imperial borderlands, particularly in Nova Scotia where Britain had returned Louisbourg to the French, Anglo-Americans feared future violence with their French, Acadian, and Indigenous neighbors. To shore up the British position in Nova Scotia, the Board of Trade—led by the Earl of Halifax—ordered the creation of the port town and naval base of Halifax, Nova Scotia, in 1749. In response to this British expansion, Acadian Catholic priest Jean-Louis Le Loutre and his Maliseet, Mi'kmaq, and Franco-Acadian allies led a bloody uprising against British authorities in what became known as Father Le Loutre's War. Between 1749 and 1755, Royal Navy captain John Rous and Nova Scotia officials continuously augmented the few Royal Navy ships in the area with several small vessels and crews from the region's provincial "sea militia."

While the Lords of the Admiralty did little to support these provincial forces, funds from the Board of Trade allowed these provincial crews to bridge communication gaps, support Anglo-American infantry forces campaigning against Le Loutre's forces, and helped prevent smuggling on contested waterways around Nova Scotia. W. A. B. Douglas contends that this provincial naval force's *petite guerre* against Franco-Indigenous forces paved the way for future larger Royal Navy campaigns on the northern borderlands during the Seven Years' War.[1]

Despite successful coordination between a few Royal Navy vessels and Nova Scotia's provincial navy during the fight with Le Loutre's forces, substantial Royal Navy involvement would be necessary to secure Britain's ever-loose foothold on its northern American peripheries. This was especially true considering the French government's reestablishment of Louisbourg as a major military base after 1749. After Captain Rous seized French vessels accused of smuggling in 1751, both British and French authorities began to send large frigates each year to Halifax and Louisbourg to compete for naval hegemony in northern waters.[2] Ultimately, mutual military escalation in Nova Scotia echoed both empires' larger territorial fights elsewhere in North America, including the vast swath of land between the Ohio River and the easternmost Great Lakes.[3] By the mid-1750s, with ongoing territorial disputes and active border wars, war with France was inevitable.

Even as the stage was set for renewed imperial struggle with France, a battle raged within the British government itself over the proper role of the Royal Navy in society. As we have seen, by the end of the War of Jenkins' Ear in the late 1740s, Great Britain's new Lords of the Admiralty—including Bedford, Anson, and Sandwich—had initiated dramatic administrative reforms that would transform the Royal Navy into a hegemonic and disciplined fighting force for the rest of the eighteenth century and beyond. However, the Admiralty's centralization program faced one immediate challenge. While having jurisdiction over its own personnel, the Admiralty did not determine the empire's general naval policy itself and followed orders from the king's cabinet ministers such as the First Lord of the Treasury (the period's equivalent of the prime minister) and various secretaries of state, including the secretary of state for the Southern Department, who oversaw the empire's Western European and American affairs.

In the interwar years between 1748 and 1754, the parsimonious Henry Pelham dominated the ministry and, much to the chagrin of naval reformers, called for military spending reductions and favored diplomatic solutions in foreign affairs.[4] During Pelham's tenure, French king Louis XV's government massively expanded its battle fleet while British authorities failed to refit decaying ships or to keep up the pace with their archrival. Even though the British fleet would eventually catch up with their foes, naval unpreparedness would create several logistical problems at the beginning of the Seven Years' War in 1756.[5] By the 1750s, however, ministerial and strategic reforms in London inspired by the new Pitt administration would lead to more effective deployment of naval assets for the remainder of the war and contribute to Britain's ultimate victory in 1763.

With previous experience of inadequate Royal Navy assistance in mind, Anglo-Americans throughout North American began to fit out provincial navies upon the outbreak of war with France. However, by the end of the 1750s, the Royal Navy's increasingly aggressive and successful campaigning in the New World made these provincial efforts unnecessary for the first time in a century. During the final years of the Seven Years' War, provincial governments felt comfortable in passing the responsibility of coastal defense to the Royal Navy. While the Royal Navy's novel North American expansion and campaigning was welcomed by colonists, its continued reliance on impressment and aggressive enforcement of postwar imperial trade policies would ultimately anger American dissidents and pave the way for the imperial crisis of the 1760s and 1770s.

PROVINCIAL NAVAL PLANNING AND THE DECLINE OF PROVINCIAL NAVAL OPERATIONS, C. 1754–1758

Throughout the Seven Years' War, Anglo-American governments from Nova Scotia to Barbados deployed a few provincial ships to assist Royal Navy forces and to defend their own shores when royal ships were far away or incapacitated. For the first time, this cooperation even extended to a joint provincial–Royal Navy fleet on the Great Lakes. Despite initial expectations that colonial governments would have to contribute large provincial naval forces to support the imperial war effort as they had done in previous conflicts, the Royal Navy's expanding presence and naval supremacy after 1758 made the existence of extensive provincial navies unnecessary.

By 1754, the British government and Anglo-American governments faced the dual crisis of French military expansion and increasingly strained relations with their traditional Iroquois allies. To solidify the Anglo-American partnership with the Iroquois as war clouds loomed and to facilitate defense plans, the Earl of Halifax and the Board of Trade called on the northern colonial governments to hold a joint conference at Albany, New York, that summer.[6] Historians throughout the last two centuries have frequently cited some of the conference participants' calls for a general colonial political union as early birth pangs of the future United States. More recently, however, scholars such as Andrew D. M. Beaumont have made the case that both British and Anglo-American authorities were equally eager to create an organized American political union for mutual military assistance.[7]

While scholars may disagree over connections between the various plans for colonial union at the Albany Congress and the future American Revolution, they have seldom noted the importance of one precocious element of these proposals: an early drive for a multicolony naval force of some kind to contest the French on the Great Lakes and the Atlantic. The most visionary naval plan at Albany came from Pennsylvania delegate Benjamin Franklin. Franklin had previously defied Philadelphia's Quaker elite when he campaigned for the commissioning of a provincial naval warship to guard the colony from French raids in 1747 at the end of the War of Jenkins' Ear. Expanding on his proposal from seven years before, Franklin suggested that a prospective American grand council and congress—under the authority of the British government, of course—would fund and construct "guard-vessels to scour the coasts from privateers in time of war, and protect the trade."

In his defense of the final Albany Plan, Franklin argued that "small vessels of force are sometimes necessary in the colonies to scour the coast of small privateers. These being provided by the Union, will be an advantage in turn to the colonies which are situated on the sea, and whose frontiers on the land-side, being coverd by other colonies, reap but little immediate benefit from the advanced forts."[8] While other delegates, including Thomas Pownall—an unofficial representative of the Earl of Halifax at the conference and future governor of Massachusetts—made vague arguments for a provincial naval force on the Great Lakes and seacoast, Franklin's proposal—which would ultimately be the basis for the final draft of the Albany Congress's Plan of Union—was the only plan that called for a centralized intercolonial navy.[9] This plan clearly demonstrated the fact that Anglo-American leaders expected that they would need substantial provincial naval forces in the coming fight with the French.

Despite his plan's novelty, Franklin's proposal for a proto–Continental Navy was premature. On the one hand, the Albany Congress itself met with little support from colonial legislatures who prized local autonomy over a united colonial military alliance.[10] On the other hand, imperial officials seemed to be just as disinterested in a major colonial maritime force as their Anglo-American constituents. After all, the Board of Trade's own simultaneous proposal for a colonial union omitted discussions of naval defense. Additionally, for reasons that are unclear, the board reported to the king that Albany commissioners had planned a "Naval establishment upon the Lake to secure the navigation" but did not mention Franklin's calls for provincial guard ships on the seas as well.[11] While Halifax's board was not opposed to supporting provincial navies (as evidenced by their support for Nova Scotia's "sea militia" throughout the interwar period), there was clearly little interest in a multicolony provincial blue water force.

Even though the plans for colonial union came to naught, one element from the discussions survived: imperial support for a naval force on the Great Lakes. In 1755, the Duke of Newcastle ordered British Army general Edward Braddock to take charge of all land-based military operations in the colonies. By April, Braddock met with the governors of Massachusetts, New York, Pennsylvania, Maryland, and Virginia in Alexandria, Virginia, to coordinate war plans. Braddock informed the governors that imperial officials had called for a multipronged attack on both French frontier forts as well as French Canada itself.[12] The various governors agreed with Braddock that a naval force was necessary on New York's contested borderlands and "advised the building of two Vessels of Sixty Tons upon the Lake Ontario . . . according to a Draught to be sent By [Royal Navy] Commodore Keppell, who desired that an Account might be laid before him of the Cost of 'em, and undertook to defray it." The attendees delegated Massachusetts governor William Shirley—a man with substantial provincial naval experience himself—with the task of coordinating the lake fleet. The attendees also planned for similar vessels to be built at Lake Erie, with the expenses of naval and land defenses there to be covered by the provincial governments of Virginia, Maryland, and Pennsylvania.[13]

The plans at the Alexandria, Virginia, meeting between Braddock and the colonial governors reflected larger British and French concerns over naval mastery of the Great Lakes—a goal that both imperial rivals saw as key to gaining mastery over the continent itself. While imperial officials ordered the construction of lake warships to counter the French fleet on Lake Ontario and elsewhere, lake crews on small craft typically avoided large fights with enemy vessels. Instead, they transported troops and supplies, scouted for enemy advances, and attempted to intercept enemy lines of communication.[14] These vessels involved some personnel from the Royal Navy but largely fell under the British Army's aegis.

The Lake Ontario navy of 1755–56 was a rare example of a fusion between the Royal Navy and colonial provincial naval forces. While Commodore Keppel (and by extension, the Crown) paid the wages of the fleet's predominantly Anglo-American sailors, provincial governments themselves largely financed the construction of the seven-vessel fleet on Lake Ontario. Even though Royal Navy captain Housman Broadley served at the small flotilla's "commodore," Governor Shirley—acting as temporary commander of Anglo-American forces—hired merchant captains to act as Broadley's subordinate officers. In late 1755 Shirley even convinced a council of fellow Anglo-American governors to underwrite the expansion of the Lake Ontario fleet without any assistance from the Royal Navy when French naval expansion seemed imminent.

This fusion of provincial and royal naval resources extended to the Lake George–Lake Champlain theater as well. Captain Joshua Loring, a former Massachusetts privateer who had joined the Royal Navy, organized a similar fleet while British forces laid siege to Fort Carillon—later known as Fort Ticonderoga. Ultimately, Loring drew on both provincial and royal financial assistance to construct vessels such as the twenty-gun brig *Duke of Cumberland*. With a dearth of sailors in this sector of the continent, many of his crewmen were officers and soldiers drawn from provincial and regular Anglo-American and British infantry regiments.[15]

While these lake navies were certainly novel for the region, Massachusetts-based Lieutenant Colonel John Bradstreet's Batteaux Service in that same theater of operations reflected some of the southern colonies' scout boat services. Batteaux were flat, shallow-water cargo vessels that had a lengthy prewar service history on New York's inland waterways and lakes. Bradstreet's flotilla of several thousand provincial batteau men, which has been characterized by one historian as the "contemporary sister organization" to the famed Roger's Rangers, transported soldiers and supplies to frontier outposts such as Fort Oswego on Lake Ontario. Occasionally, these backwoods sailors even disembarked to fend off large groups of Franco-Indigenous raiders.[16]

Convinced by the success of these hardy mariners, in late 1757, Bradstreet—in much the same manner as Franklin's proposed colonial navy—asked the British government to finance an even more extensive multicolony batteau service led by American officers and bankrolled by imperial funds. While the British commander of North America at the time, Lord Loudoun, did not accept the petition in its full form, he did promise Crown reimbursement for the financial costs of Bradstreet's proposed 1758 naval assault on Fort Frontenac.[17] This promise reflected the fact that British Army commanders frequently drew on imperial funds to support lake navies, though various issues with credit would create issues for this system.[18] Nevertheless, by the end of the conflict in 1763, Anglo-American and imperial officials had cooperated to construct or purchase nearly thirty small war and cargo vessels on Lakes George, Ontario, Champlain, and Erie.[19]

While provincial and royal governments pooled resources to fit out lake fleets, no such cooperation existed in American and West Indian port cities, where the costs of coastal provincial navies were as high as ever. Between 1756 and 1758, numerous provincial governments fitted out (or planned to fit out) seagoing guard vessels to combat the ever-present threat of French privateers. This naval effort was understandably more potent in regions directly affected by warfare with the French, particularly in New England. For instance, after news of the 1756 declaration of war against France, Governor Shirley's administration in Massachusetts spearheaded an effort to use local tax money to fund two provincial warships: the snow, *Prince of Wales*, captained by Nathaniel Dowse; and the frigate *King George*, captained by Benjamin Hallowell Jr.

The aptly named Massachusetts frigate *King George* was a particularly useful adjunct for Royal Navy forces operating in northern waters throughout the Seven Years' War.[20] In the summer of 1757, Lord Halifax's ally and Shirley's successor as governor, Thomas Pownall, reported to Prime Minister William Pitt that Massachusetts had a "naval Establishment (which no other Province has)." While Pownall apparently did not realize that most British colonies had established provincial navies at some point throughout their history, he did realize their utility and expressed the hope that they would be useful assistants to the Royal Navy.[21] Indeed, near the end of the conflict in October 1762, the next Massachusetts governor, Francis Bernard, bragged that the *King George* was instrumental to Royal Navy admiral Lord Colville's victory over the French in Newfoundland. Even though the "junction of the *King George* with Lord Collville appeared to be an accidental meeting instead of a Concerted Measure . . . I received from Lord Colville such an high testimony of Captain Hollowel [sic]."[22]

While provincial governors of Massachusetts envisioned the colony's navy as a useful adjunct for an ever-expanding Royal Navy presence in the North Atlantic, colonies to its immediate south and in the West Indies deployed provincial naval forces on a much more ad-hoc basis in order to stop the growing threat of French privateers. Even though the Royal Navy effectively eliminated the French navy as a serious threat by the late 1750s, it was unable to fully control elusive French private men of war until the end of the war. By the early 1760s, French privateers had captured nearly 1,400 British ships in the Caribbean theater alone.[23]

In 1757, the Connecticut government purchased a brigantine, *Tartar*, and assigned Michael Burnham as its captain. Burnham, who had been one of the last captains of the colony sloop *Defence* during the previous imperial war, led his crew on a journey to the West Indies to protect Connecticut trade interests there. By 1758, without any significant debate, the provincial government decided to sell *Tartar*.[24] Around the same time, Rhode Island's government took charge of two vessels—including the privateer brigantine *Abercrombie*—and ordered them to hunt for a French privateer who was harassing English commerce off of Block Island. While Rhode Island's government commissioned numerous privateers throughout the conflict, it abandoned its only attempt at a provincially funded guard vessel in late 1758.[25]

Farther south in 1757, the traditionally pacifist Pennsylvania government agreed to fit out a twenty-two-gun provincial vessel known as *Pennsylvania Frigate*, with the express purpose of the "Protection of our Trade."[26] Far from answering this purpose, *Pennsylvania Frigate*'s captain John Sibbald faced accusations of inaction and cowardice in colonial newspapers in New York and Pennsylvania. By late 1758, the Pennsylvania House of Representatives Committee of Correspondence wrote the Lords of the Admiralty to complain about the "Losses sustained by the Merchants of this colony . . . notwithstanding the great Expence they have for some Time past been at in supporting a Ship of War to guard the Coast, and humbly pray the Assistance from our Mother Country, of a Vessel or Vessels of superior Force."[27] Despite early disappointment with the frigate, the Pennsylvania government kept the vessel cruising to protect the colony's trade for the rest of the conflict. Pennsylvania would prove to be the only colony other than Massachusetts to keep a provincial frigate cruising for this long in the war.[28]

Provincial naval patrols also occurred in the southern colonies, although the immediate French threat was much more muted there. Furthermore, Spain's late entry and brief participation in the war in the year 1762 alone made any significant campaigns against St. Augustine unnecessary. Nevertheless, these factors did not prevent provincial leaders in South Carolina and Georgia from utilizing provincial naval forces at times. In South Carolina, a committee of concerned merchants and planters discussed creating a voluntary fund from which "one or two Vessels of War may upon any sudden Occasion be immediately fitted out" in case French privateers were to attack Charles Town. They contended that "his Majesty's Ships cannot at all Times go over the Bar, the Consequences of which we need not mention."[29]

Even though locals continued to manage their own coastal defenses when needed, an episode in the summer of 1757 demonstrated just how intermeshed provincial and royal defense efforts had become. When a French privateer attacked local merchant vessels in the waters of Charles Town, the local government fitted out an emergency fleet of two small vessels to pursue it. While one provincial vessel had a crew of local volunteers and infantrymen from Lieutenant Colonel Henry Bouquet's Sixtieth Regiment of Foot (the Royal Americans), the other provincial vessel was manned entirely by Royal Navy sailors and marines from HMS *Arundel*.[30]

Even though the Royal Navy and elements of the British Army demonstrated a willingness to assist South Carolina's provincial forces in 1757, this cooperation seems to have ended by 1758. A Charles Town correspondent reported that even "Tho' we have not a Man of War or other Vessel cruizing from Port in this Province, to protect our Coasts against the Insults of the French Privateers that may be upon it, we are assured that the Province of Georgia has—a fine Sloop having been impressed there." Georgia governor Henry Ellis put the vessel *Tryal* under the command of a privateer captain, assigned the captain of the colony's scout boat as a pilot, and added the sailors from the scout boat as well as several volunteers to its crew. The source reported that in a subsequent battle with a French privateer crew (that included escaped enslaved Africans from South Carolina), *Tryal*'s crew suffered many casualties but successfully withstood several boarding attempts. With this pyrrhic victory in mind, the South Carolina correspondent declared that: "GEORGIA has made its Effort; and surely it must now be our Turn! 'Tis true, the Event of our Sister-Colony's Endeavours carries some Disappointment in it, but we cannot think they have been fruitless . . . those who go with a sincere Intention of finding the Enemy, seldom fail to meet them."[31] It is likely that the correspondent's conclusion was an acerbic commentary on what he believed to be the Royal Navy's alleged inactivity in patrolling for French privateers.

While it is tempting to see Ellis's provincial navy as an example of colonial self-reliance in the wake of traditional Royal Navy negligence, major changes were occurring at Whitehall that would ensure a stronger Royal Navy presence in the New World. Even independent provincial naval expeditions during the late 1750s should be seen within the context of Britain's growing military strength throughout the Atlantic world. For instance, historian Andrew D. M. Beaumont has argued that Governor Ellis's ability to "act decisively upon his own initiative" was precisely why the Board of Trade's Lord Halifax had made him the governor of Georgia. Beaumont contends that throughout 1756 and 1757, ministerial infighting and military inaction by British commanders such as Lord Loudoun damaged Britain's war effort. To counter this, Lord Halifax depended upon the colonial governors to carry on the fight against the French with local resources. In short, Beaumont holds that even colonial authorities acting on their own initiative could still advance metropolitan military goals.³² By 1759, provincial naval forces in South Carolina and Georgia had made some strides against French privateers, but these minor naval forces paled in comparison to the large southern provincial navies of the War of Jenkins' Ear.

Just as in the last war, provincial navies were only moderately active in the well-patrolled and lucrative West Indian and Atlantic island colonies. In fact, within the first few years of the war, political squabbling, arguments over finance, and the overwhelming presence of Royal Navy guardships limited the service lives of even those few provincial ships in the region.³³ That is not to say that there was never any use for local defense vessels in island provinces. For instance, late in the war in 1761, the Bermuda government fitted out two sloops to chase after French privateer sloops; the sloops successfully chased the raiders away.³⁴

Despite occasional utility for emergency fleets and provincial guardships, the Jamaica governor's 1757 speech to a joint session of his council and the island's assembly provides a poignant picture of the decline of provincial navies in the West Indies: "I apprehend there will be no Occasion for an Island Sloop, two [royal] Vessels having been already commissioned, by an Order from the Lords of the Admiralty, for the immediate Protection of our Coasts[;] The great Sums of Money usually expended in Time of War for that Service will now be saved, the Country relieved from so heavy a Burthen, and the purpose more fully answered."³⁵

The decline in provincial naval warfare throughout the Atlantic world in the late 1750s coincided with the decline of its sister institution: privateering. Though Parliament initially encouraged widespread privateering at the beginning of the conflict, and even though it allowed privateers to raid neutral merchant ships that carried French goods, by 1759 the British government decided to limit the issue of Letters of Marque when tensions arose with neutral powers concerned about British assaults on their shipping. Aside from diplomatic concerns over privateering excesses, Royal Navy vessels were also more efficient in commerce raiding during the Seven Years' War than privateers. In his study on British privateers throughout the eighteenth century, historian David Starkey calculated that between 1739 and 1751, British authorities condemned 408 enemy vessels captured by privateers and 449 vessels captured by the Royal Navy. Between 1756 and 1763, British courts condemned 382 privateer prizes and 794 Royal Navy prizes. Starkey connects the decline in British privateering to the Royal Navy's successful campaigns in the second half of the Seven Years' War.³⁶

This decline in privateering after 1759 was noticeable in America as well. New York merchants had fitted out three times as many privateers in the Seven Years' War as they had in the War of Jenkins' Ear. By 1759, however, overhunting of enemy commerce reduced the number of prizes available for privateers. New York's Vice Admiralty judge Lewis Morris—a veteran administrator that had overseen privateering cases since the War of Jenkins' Ear—condemned more prizes in 1758 than in any year in his long career but saw fewer and fewer cases as British victories and "overfishing" of French prizes continued.[37] All in all, Royal Navy military victories throughout the Atlantic world by the end of the decade disincentivized provincial governments from outfitting large provincial navies and private merchants from pursuing privateering.

Royal Navy Supremacy at Sea, 1758–1763

Even though provincial governments initially sought to fit out provincial navies in the war's early years, the Royal Navy's expansion and victories over its French enemies after 1758 made the creation of extensive provincial navies unnecessary. To fully appreciate why colonial governments consciously decided to cut back on provincial navy spending, one must fully examine the reasons behind the Royal Navy's growing hegemony in the Atlantic world in the late 1750s.

The seeds of Royal Navy squadrons replacing private naval squadrons at sea were sown as early as the mid-1750s, but imperial naval strength would only fully be realized by the end of the decade. W. A. B. Douglas argues that when Royal Navy captain John Rous led a force of royal ships to help take French Fort Beausejour in Nova Scotia in the summer of 1755, it was "perhaps symptomatic of that state of affairs that [Rous's squadron was] composed entirely of King's ships rather than a mixed force of provincial and [royal] naval vessels."[38]

Broadly speaking, the Royal Navy's ultimate maritime hegemony by the end of the 1750s did not come easily. The Duke of Newcastle, Sir Thomas Pelham-Holles (the brother of Prime Minister Henry Pelham and his successor after 1756), initially hoped to contain French aggression to North America and to avoid an all-out European conflagration. After all, his administration feared that if another worldwide imperial war broke out, the battle fleets of the recently expanded French and Spanish navies would potentially outnumber and overpower the Royal Navy. While important members of the Whig opposition such as William Pitt—the secretary of state for the Southern Department—and naval reformers in the Admiralty such as Lords Bedford and Sandwich called for an aggressive maritime assault on France, Newcastle and his allies insisted that diplomatic solutions in Europe and limited warfare against the French were the best methods to achieve victory. Ultimately, the ministry did not dedicate adequate naval forces to intercept French reinforcements sailing for the New World—a misstep that played a major role in the expansion of the war beyond the North American continent in 1756.[39]

What had begun as border skirmishes in North America quickly became a major world war between 1756 and 1758. In the English Channel and on the French coast, the Royal Navy's Western Squadron failed to stop three major French squadrons sailing for the Caribbean and Canada. These fleets reinforced Fortress Louisbourg—an event that delayed Anglo-American plans for an assault on French Canada. To calm public anger over mediocre progress against French forces, Pitt planned for a major Royal Navy–British Army assault on the port of Rochefort, France, in late 1757. Unfortunately for Pitt, interservice rivalries, faulty intelligence, and bad weather forced the invasion force to withdraw. A subsequent joint raid on the French port of St. Malo in the summer of 1758 was more successful and resulted in the destruction of eighty French privateer vessels and merchant ships along with four French naval vessels under construction.[40]

In the Mediterranean, the French captured the island of Minorca in the spring of 1756. A Royal Navy relief force led by Admiral Byng failed to recover the island, and Byng himself was executed by the Admiralty for his alleged inaction. Despite this loss, however, by 1758 the Royal Navy was able to contain the main French fleet in the Mediterranean and prevented it from sailing across the Atlantic to reinforce Louisbourg.[41] Around the same time, Royal Navy forces also seized French Senegal in West Africa.[42]

In the East Indies, the French and British East India Companies, along with their respective Indian allies, had long been at war with one another. A Royal Navy force had already been sent to India in 1755 to assist the East India Company's own naval forces (the Bombay Marine) in a fight against their enemies in the Angrian Indian kingdom and was prepared for the larger fight against the French when news of war arrived in 1756. Between 1756 and 1759, British and French forces—alongside their Indian partners—fought a largely inconclusive war of attrition. Both British and French squadrons fought each other to a standstill on numerous occasions, and both sides captured important trading outposts from one another. By 1761, however, British forces had largely forced the French out of India with the capture of Pondicherry.[43]

While clashes between the British and French empires occurred throughout the world, William Pitt's tenure as prime minister after 1756 led to a sea change in British war planning. With growing resentment against long-standing Whig concerns with political and military involvement in continental European affairs, Pitt and his ministerial allies decided to focus on a "Blue Water" vision of British empire. This vision utilized naval power to expand the British imperial reach into the Americas—a move that would both bolster British wealth and harm the empire's French enemies.[44]

The positive effects of this renewed British attention to the North American theater became apparent in the summer of 1758 when Admiral Boscawen led over twenty-one ships-of-the-line and two frigates—the first Royal Navy fleet that had ever wintered in Nova Scotia—alongside twelve thousand soldiers in a successful assault on Fortress Louisbourg. This act would serve as the first step in the larger conquest of Canada that would occur throughout the next two years. It is telling that Massachusetts's government, which had spearheaded a large provincial flotilla in the siege of 1745, added little more to this campaign than to utilize the provincial frigate *King George* as a scouting vessel and commerce raider on the coast surrounding Louisbourg.[45] Ultimately, the capture of Louisbourg demonstrated that the Blue Water strategy hinged more on royal than provincial naval resources.

Even though fighting continued in every corner of the world, the Royal Navy held the advantage over the French in North America and in the West Indies after the 1758 seizure of Louisbourg and subsequent 1759 capture of Quebec. While Pitt's adoption of more aggressive naval warfare against French interests certainly aided the British war effort, one other factor worked in the British Empire's favor: Spain's continued neutrality. King Filip VI of Spain had long sought to maintain peaceful relations with the British Empire. However, when the king died in the summer of 1759, the Spanish government's devotion to neutrality died with him. His half-brother and successor, Charles III, sought to restore his kingdom's traditional alliance with France and oversaw a further expansion of the Spanish Navy and sent squadrons to reinforce garrisons in the West Indies.

At the same time, ongoing peace talks between Britain and France proved to be unfruitful. Pitt's desire to continue the war at any cost and to welcome a fight with the increasingly belligerent Spanish was unpopular to the war-weary British public, and the prime minister resigned in the autumn of 1761. Despite his resignation, Spain belatedly formalized a military alliance with France in the winter of 1761. Within weeks of the declaration of war, officials in London planned to use the large British infantry and naval forces already campaigning against the French in the Caribbean—along with various Anglo-American provincial forces and units of free volunteers of color—to capture Havana, Cuba. The British capture of the city in the summer of 1762 would strengthen British diplomatic efforts in ongoing peace negotiations in Europe.[46]

It is interesting to note that southern provincial governments did not engage in any major military campaigns against their traditional Spanish foes. While Anglo-Americans undoubtedly fitted out privateers and occasional provincial fleets, a policy of proactive defense at this late stage in the conflict was preferred over major campaigns against the Spanish. This sentiment was best expressed in a spring of 1762 issue of the *South Carolina Gazette*, which reported that "The general assembly of this province have [sic] resolved to continue both the scout-boats, and the look-outs, during the continuation of the present war with France and Spain."[47] Maintaining scout boats was a far cry from planning a major invasion of Spanish Florida. This relative inactivity can only be attributed to one thing: the royal military's successful campaigns against France and Spain.

All told, the Royal Navy played a key role in achieving Britain's first imperial victory over its foes in half a century. In London, the Admiralty Board's continued insistence on professionalization, aggressive strategies, new battleship designs, the capture of large numbers of enemy mariners, British assaults on neutral ships carrying French goods, and French economic collapse all contributed to Britain's growing naval advantage as the long war blazed on. Behind aggressive military expansion throughout the war, new methods for distributing supplies to ships, and Whig prime minister Pitt's ability to get consistent credit and funding for both the navy and army also played large roles in the British naval victory.[48]

The Admiralty's growing interest in overseas conquests paralleled an even larger shift in the British government's relationship with its overseas empire in America. Throughout the last several decades, a number of scholars have noted an increased metropolitan interest in colonial military defense with the onset of the Seven Years' War. Kurt Nagel has argued that by the late 1740s, the British government had begun to take the reins of colonial military defense policies while also continuing to insist on colonial self-defense measures—a contradiction that would play a role in fostering the imperial crisis of the 1760s. London's total involvement in colonial military affairs would crystallize by the Seven Years' War with Pitt's aforementioned adoption of a Blue Water strategy that increased military involvement in America.⁴⁹ Thus, the Royal Navy's growing presence in American naval warfare was symptomatic of the growing metropolitan interest in the governance and defense of its American colonies by the 1760s.

Even if the Royal Navy's expanded operations in the Atlantic world supplanted the need for provincial navies, British naval dominance over the French and Spanish required significant American assistance and sacrifice. For instance, in 1759, after Royal Navy admiral Durell requested men for his ever-undermanned squadron at Halifax, Nova Scotia, Rhode Island's government promised bounties "out of the general treasury, over and above the King's, of forty shillings sterling" for all men who would join the royal fleet.⁵⁰ This sort of spending did not come easily. For instance, even though Parliament had reimbursed *some* of Massachusetts's war debts by 1759, the colony continued to employ more than 25 percent of its adult male population in infantry and naval services (including on batteau boats, privateers, and on the frigate *King George*) with its own coin. Even though these charges sapped the colony's economic strength in the short term, the promise of future parliamentary reimbursement encouraged Massachusetts and other colonial governments to provide thousands of soldiers and sailors for the imperial cause.

While economic reimbursement from London was vital in securing Anglo-American military expenditures throughout the conflict, another force also drove colonial governments to continue to support the imperial cause: a growing belief among Anglo-Americans that they were equal "partners" with the British military in the war against the French. Scholars have noted that this growing patriotic fervor was not shared by British Army leaders (particularly General Amherst), who saw colonial governments and their forces as fickle subordinates rather than as imperial partners.⁵¹ While these tensions had existed on land for some time, they also continued to plague provincial-royal cooperation at sea. As discussed in the previous chapter, Anglo-Americans had failed to convince imperial authorities that their provincial warships at the 1745 Siege of Louisbourg had been the equals of Royal Navy frigates.

As ever, the sharpest controversy between Anglo-Americans and the Royal Navy was impressment. In the aftermath of the 1747 Knowles Riots, Royal Navy captains typically only impressed American sailors already at sea rather than in port. This new strategy did little to assuage colonial authorities and met with violent resistance near Boston Harbor in 1758 when a merchant vessel fired on a boat carrying a press gang from HMS *Hunter*.⁵² Despite early reservations, the Royal Navy did not always limit its impressment to sea—a fact that would bring about significant resistance throughout the northern port cities.

Continued drama over impressment policies during the Seven Years' War can only be understood within the context of other controversial recruitment policies by Crown forces. Throughout colonial America's towns and cities, British Army recruiters had begun to rely on coercion and violence to force Anglo-Americans into personnel-depleted regiments. Thus, Anglo-American mobs frequently assaulted regular army recruiting parties; mob violence against Royal Navy press gangs was a simultaneous occurrence.[53] For New Yorkers, these two threats coalesced in the spring of 1757 when Lord Loudon's infantry forces assisted the Royal Navy in impressing over eight hundred men.

While mariners could not resist Royal Navy press gangs that had large, red-coated units at their disposal, individual crews did put up hefty resistance when they had the chance. One representative example occurred in 1760, when the crew of the privateer *Sampson* engaged in a vicious firefight with a Royal Navy crew that tried to impress them. Even though local lawmen tried to help the press gang, the privateer crew was largely able to escape unscathed from New York Harbor.[54]

Of course, this resistance was not limited to Royal Navy impressment efforts. For instance, in 1758, a number of sailors fired on a New York militia company that tried to force them into the colony's transport service.[55] While violent resistance to provincial impressment efforts was not unheard of, sailors were often far more concerned with the much more expansive threat of Royal Navy impressment.

Although colonists resented the Royal Navy for its impressment policy, they also grew to despise its role in suppressing illicit Anglo-American trade with the French. As early as 1755, Admiralty officials took note of the widespread North American trade with the French in Louisbourg and the West Indies, and ordered Admiral Augustus Keppel to patrol for smugglers. Anglo-American smugglers often had patronage from colonial governors, including Pennsylvania governor William Denny. Denny sold "flag-of-truce" passes to merchant captains who would tacitly go on diplomatic missions to French territories with the understanding that they would engage in illicit trading with the enemy.

Metropolitan anger over this smuggling coupled with provincial anger over British attempts to end the practice further strained relations between periphery and center. When British commanders such as Lord Loudon placed embargoes on colonial ports to prevent this trade, Anglo-Americans loudly protested over their financial losses. By the early 1760s, parliamentary anger at this widespread trade would prove to be fundamental in its decision to curb provincial autonomy with numerous imperial reforms, beginning with legislation surrounding Writs of Assistance (which will be discussed in the next chapter).[56]

By the early 1760s, imperial planners were becoming convinced that they would need to permanently expand the British military presence in North America to both defend colonists from Britain's traditional foes *and* to enforce their policies within the colonies. To fund permanent wartime garrisons—and to ensure provincial dependence on Westminster—the British government began to plan various major reforms, including expanding vice admiralty court jurisdiction and enforcing customs laws with a fleet of purpose-built coast guard vessels. Historian Eliga Gould argues that Britain's decision to strenuously enforce trade laws and tax policies began to make enemies of important merchants and ordinary sailors who had "cut their political teeth resisting the navy's wartime press gangs during the 1740s and 1750s."[57] As we will see in the next chapter, it is ironic that those imperial laws designed to bring Anglo-Americans closer in line with British maritime policies had the effect of alienating the colonies' sailors to the point of all-out rebellion.

Between 1754 and 1757, provincial governments planned for major provincial naval campaigns against the French as they had in the previous imperial conflict. However, beginning with the capture of Louisbourg in 1758, Pitt's Blue Water strategy, which relied on the Royal Navy to spearhead the conquest of French possessions in North America, made the existence of large provincial navies unnecessary. With dozens of Royal Navy warships actively pursuing French privateers and capturing French ports, provincial governments felt that they could finally delegate the responsibilities of coastal defense to their imperial overlords. While an expanded Royal Navy presence may have made Anglo-Americans feel secure while the war raged on, the imperial fleet would play an important role in exacerbating the postwar imperial crisis of the 1760s and 1770s that would ultimately pave the way for the Revolutionary War.

6

The Legacy of Provincial Navies and the Navies of the American Revolution, c. 1762–1775

At the beginning of the 1760s, the Seven Years' War raged on, and Spain belatedly joined the fight as an eleventh-hour entrant to the global conflict. Despite Spain's entry into the war on the side of the French, the momentum of the conflict had already significantly swung in favor of British and Anglo-American forces on land and sea. The Royal Navy—perhaps for the first time in the century and a half of English colonization in the Americas—could truly claim complete and total maritime supremacy throughout the Atlantic basin.

Nevertheless, even as the British Empire was poised for total victory in this final war of the pre-Revolutionary era, those latent metropolitan-provincial tensions mentioned in the preceding chapters—namely Anglo-American anger over Royal Navy impressment and a legacy of insufficient naval support from the British government—would metastasize during the final months of the Seven Years' War and throughout the imperial crisis of the 1760s–70s. While historians have long understood the largest cause of the Revolutionary War to be Anglo-American anger at British taxation without representation (largely designed to cover the costs of imperial defense), scant attention has been paid to the ways in which the legacy of colonial provincial warfare shaped American resistance to imperial authorities during this Revolutionary moment. Ultimately, more than a century of independent naval defense coupled with latent provincial antipathy to Royal Navy forces would guide initial Anglo-American resistance to British authorities at sea. In the opening years of the War for Independence, these very same experiences would play fundamental roles in the creation of the nation's first naval forces.

Provincial Naval Echoes and the Imperial Crisis, 1762–1775

Unsurprisingly, the first major connection between pre-Revolutionary provincial navies and the imperial crisis occurred in the ever-turbulent port city of Boston. By the middle of 1760, Boston, like many other northeastern ports, faced an economic recession as the war with France began to wind down. It also faced political infighting between conservative elites (and supporters of extending Governor Francis Bernard's prerogative powers) such as the colony's chief justice Thomas Hutchinson and populist politicians such as James Otis. Among the sharpest disputes that arose between these political factions was the battle over Writs of Assistance, one of the first controversial imperial reforms of the 1760s. The British government had equipped customs officers with greater authority to utilize search warrants (writs) on vessels suspected of smuggling. Otis's campaign against the Writs of Assistance and Hutchinson's attempts to ban the popular Boston town meeting endeared him to common laborers and merchants concerned with increasing imperial trade restrictions.[1]

Aside from Otis's resistance to what he considered growing imperial overreach, he also opposed Governor Bernard's handling of the colony's provincial navy in the final year of the war against France—a case he documented in the 1762 pamphlet *A Vindication of the Conduct of the House of Representatives of the Province of the Massachusetts Bay*. One early nineteenth-century historian made the bold case that this pamphlet "has been considered the original source, from which all subsequent arguments against taxation were derived."[2] With the French having commenced an assault on Newfoundland that summer and with coastal fishermen fearing a renewed assault on the New England fisheries, Bernard and his council expanded the crew of the provincial sloop *Massachusetts* and sent it on various patrols without consulting the colony's assembly. Bernard's unilateral strategy may have appeared harmless to the governor, but what may have seemed like a small quibble over the defense of the coast set the pace for larger constitutional arguments that would resound throughout the next few decades.

Overall, Otis's larger argument was not with the existence of a provincial navy in Massachusetts, though he questioned if "the province's trade has truly received a Benefit from those Vessels equal to the Tax . . . paid for their Support." Rather, Otis characterized Bernard's fitting out of the sloop without approaching the assembly, coupled with other extraparliamentary expenditures, as a symptom of arbitrary executive power. A legislative committee responded with the claim that "No Necessity therefore can be sufficient to justify a house of Representatives in giving up such a Priviledge; for it would be of little consequence to the people whether they were subject to ... the King of Great Britain or the French King, if both were arbitrary, as both would be if both could levy Taxes without Parliament."[3] Ultimately, as in so many cases throughout the Atlantic world in the preceding century, battles over provincial navies reflected larger sociopolitical divisions and tensions within colonial society rather than squabbles over naval policy.

While Otis and his colleagues equated Bernard's naval expenditure with taxing the populace without representation—a clarion cry that would resound throughout colonial protests for the next decade—they also expressed an anxiety common throughout seventeenth- and eighteenth-century British politics: that an executive would keep a standing military force to arbitrarily oppress his subjects. This fear was evident when Otis wondered "If the Governor and Council can fit out one man of war, inlist men, grant a bounty and make establishments, why not for a navy, if to them it shall seem necessary, and they can make themselves the sole judges of this necessity."[4] Historian Sarah Kinkel has recently made the case that later in that decade, Anglo-Americans dissidents—like some of their Whiggish compatriots in Parliament—protested Royal Navy enforcement of metropolitan trade laws partly due to their "preexisting fears about a professional military."[5] While men like Otis did not oppose provincial navies on principle, they did fear that excessively powerful governors could wield them in the same manner as a standing army to squash the rights of the citizenry.

While it may seem hyperbolic to assume a governor could maintain a private navy to enforce his will (or imperial laws), there were some cases where provincial navies supported unpopular British policies during the imperial crisis. For instance, the Georgia scout boat *Prince George*—still active decades after General Oglethorpe created Georgia's provincial navy—was fundamental in securing the delivery of stamps after Parliament's infamous 1765 Stamp Act. This act was one of Parliament's first major attempts at external taxation on internal colonial commerce and required colonists to pay a duty on various paper goods and licenses. This wildly unpopular act met immediate resistance throughout the American colonies.

In Georgia, merchants were furious that they could not export rice (the colony's cash crop) without customs papers with the stamps affixed and took to the streets to protest the policy throughout the autumn of 1765.

Throughout the next several months, Wright mobilized the colony's Crown-funded ranger force, elements of the Royal Navy, and volunteers to defend the stamps and his own safety when the local chapter of the Sons of Liberty threatened numerous violent riots. It will be recalled that Georgia's provincial forces had been on the royal pay roster since Oglethorpe's military campaigns of the 1740s. Along with a few colonial rangers, the crew of *Prince George* transported and guarded the colony's stamp collector during his initial landing.[6] After the governor's swift response, one anonymous Georgia Whig lamented that "Our liberty here is at a very low ebb." Undoubtedly, the governor's ability to utilize a royally funded infantry force and provincial scout boat in support of the Stamp Act did little to quell colonial fears that their rights were threatened by standing military forces.[7]

While the imperial crisis raised larger questions over the ability of the governor to use provincial navies to enforce unpopular imperial mandates, the legacy of provincial naval service from previous conflicts also shaped the way Anglo-Americans protested British policies. For example, as early as 1764, a committee of Massachusetts politicians from the governor's Council and Assembly—leery of reports that the British ministry and Parliament were plotting a round of taxes on the American colonies—drew on their colony's century and a half of military service to demonstrate their loyalty to the Crown and to decry imperial taxation. In response to Parliament's reason for raising taxes on the colonists, to "defray the charges of a war undertaken for [the colonists'] defence, to which it is said they have never yet sufficiently contributed, the Province of Massachusetts Bay deem it proper briefly to set forth their own . . . exertions and expenses in the common cause."

Among the major expenses the committee delineated were the various attacks on Canada throughout the previous imperial wars, contributions to campaigns in the Seven Years' War, and the commissioning of "armed vessels for the protection of trade, [which] cost 34,795 [pounds]." While still in debt from the most recent conflict, the colony's government could still declare that "From its infancy to the present age, this colony, with no expense to the Crown, has defended the territory granted to it; and thereby mightily extended the British empire and immensely increased the British commerce."[8] While provincial naval expenses were only one factor in the colony's long list of complaints against recent British trade acts, this complaint reiterated long-held provincial anger at bearing the brunt of the costs of naval defense. In essence, New England Whigs questioned why they should have to underwrite the costs of the Seven Years' War while they had shouldered the costs of provincial defense for over a century.

While Massachusetts explicitly listed its provincial naval expenses as evidence that the colony should not be taxed by Parliament, other colonies drew on more general descriptions of their military exertions to justify their protests. For instance, in his 1764 pamphlet *The Rights of the Colonies Examined*, Rhode Island politician Stephen Hopkins—a future signer of the Declaration of Independence and founder of the Continental Navy—detailed various colonies' historical wartime sacrifices as evidence that they should not be taxed by Parliament. For instance, Hopkins argued that "in the year 1746, when the Duke D'Anville came out from France, with the most formidable French fleet that ever was in the American seas, enraged at these colonies for the loss of Louisbourg, the year before, and with orders to make an attack on them; even in this greatest exigence, these colonies were left to the protection of heaven, and their own efforts."[9]

While Hopkins made no explicit mention of his colony's naval service in this example, he must certainly have considered the fact that Royal Navy admiral Warren had personally requested Rhode Island's colony sloop *Tartar* to scout for DuCasse's squadron during the invasion scare.[10] In the mind of Anglo-American dissidents, parliamentary taxation to fund standing military forces punished colonial governments that had funded their own defense measures for generations.

Although Anglo-Americans drew on their history of provincial naval expenses and general military costs to argue against British taxation without representation, this same legacy also defined how many disaffected colonists opposed the metropole's expansive maritime enforcement policies. One critical component of Parliament's plans to levy taxes on American colonists in the years following the Seven Years' War was the Admiralty's plan to expand the North American Squadron's peacetime fleet to twenty-six vessels (and 3,290 sailors). While France's empire in North America had essentially come to an end in 1763, the British government hoped that maintaining a peacetime garrison of thousands of red-coated regulars—along with an expanded naval presence—would prevent future imperial competition over its new American territories.

Aside from military fears, imperial authorities also hoped the Royal Navy could crack down on widespread American smuggling—a potentially lucrative service that would serve immediate imperial interests and allow for peacetime prizes for Royal Navy crews. The first peacetime commander of the newly expanded squadron, Lord Colville—who had recently utilized Massachusetts's provincial warship *King George* in the attack on Newfoundland—zealously embraced his new powers to seize and confiscate illicit cargo and trading vessels.[11] By 1764, Admiralty officials funded the construction of six small sloops and schooners for the Royal Navy to use in the pursuit of North American and even occasional French smugglers.

While the Royal Navy did fund this small cutter fleet, the most effective and unpopular antismuggling vessels were "peacetime privateers" commissioned by the American Board of Customs Commissioners. Their crews lived off the proceeds of their captures and tarnished the reputation of the Royal Navy in American waters even though they were independent of the imperial fleet. For many Whiggish American traders and smugglers, Britain's new "sea guard" was little better than the *guarda costas* that had prowled their shores throughout the recent imperial conflicts.[12]

For many traders in Rhode Island who depended heavily on the molasses trade with the West Indies, Parliament's 1764 Sugar Tax and the British coast guard vessels that enforced imperial trade measures were a major threat to their livelihoods. Strict threats to the colony's commerce were met with a warlike response. This military response, particularly in the traditionally rebellious colony of Rhode Island, drew on nearly a century of commissioning "emergency fleets" to face immediate piratical and imperial threats. With ongoing Royal Navy captures of sugar smugglers and rumors of impressment plans, Rhode Islanders began to stage violent resistance to Royal Navy guard ships as early as 1764.[13] When Lieutenant Thomas Hill and the crew of the revenue schooner *St. John* tried to recruit sailors in Newport, local peer pressure and threats of violence stalled recruiting. To the locals' chagrin, Hill seized a local smuggling vessel soon thereafter.

With tensions escalating, a rumor arose that, when Royal Navy sailors went on shore to claim a deserter, they plundered a local farm; the locals had planned to fit out an armed vessel to attack *St. John*. Allegedly they were only deterred from this attack by the presence of the nearby warship HMS *Squirrel*. While some scholars have made the case that the planned attack on *St. John* was a mere rumor, it is clear in Hill's correspondence that he believed a mob had almost overtaken his vessel. Although Hill had been absent during the violence, some of his subordinate officers reported that a "mob filled a sloop full of men, and bore right down to board us." While Royal Navy firepower prevented this mob from boarding *St. John*, local gunners at a nearby fort fired at the mainsail of the royal schooner and forced it to fall back.[14] The violent battle for navigation in Rhode Island had commenced, and impromptu fitting out of warlike vessels would serve Rhode Islanders in their fight against the Royal Navy just as it had with pirates and other previous maritime foes.

Throughout the rest of the 1760s, Rhode Islanders violently resisted impressment attempts by the Royal Navy and even burned the British revenue sloop *Liberty*—a vessel that previously belonged to elite Boston smuggler and future Founding Father John Hancock.[15] Interestingly, the customs officer who initially seized *Liberty* from Hancock, Benjamin Hallowell, was the former provincial navy captain of Massachusetts's province ship *King George*. Hallowell's actions led to the plundering of his home during an ensuing riot in Boston, and his Tory leanings ultimately led him to flee to Canada when the Revolutionary War broke out. Even though riotous traders and Whigs drew on traditional provincial naval strategies to resist the British during the imperial crisis, Hallowell's example reminds us that previous service in Anglo-American provincial navies did not always correlate with resistance to British authority.[16]

While Rhode Islanders used mob violence to secure their shipping and sailors from the Royal Navy throughout the late 1760s, they transitioned to all-out naval assaults on royal revenue cutters by the early 1770s. This escalation of violence occurred after Lieutenant William Dudingston and the crew of the revenue schooner *Gaspee* seized numerous Rhode Island smugglers and brought them to the Vice Admiralty court in Massachusetts. In response to these seizures and Dudingston's refusal to show proof of his authority, Governor Wanton of Rhode Island engaged in a vicious war of letters with the lieutenant and his superior, Admiral Montagu.[17] In a letter to Wanton, Montagu claimed that *Gaspee* was stationed at Rhode Island to protect the locals from piracy and to end smuggling. He also claimed that "the people of Newport talk of fitting out an armed vessel to rescue any vessel the King's schooner may take carrying on an illicit trade. Let them be cautious what they do; for . . . any of them are taken, I will hang as pirates." Even though Wanton denied knowledge of local preparations to assault *Gaspee*, Montagu warned that any provincial mob would meet deadly force if they molested the king's ships.[18]

Even though Montagu feared an attack by a Rhode Island *vessel*, he could never have fathomed the multiboat attack that would occur against *Gaspee* in the summer of 1772. Many decades after the raid, the last survivor, Ephraim Bowen, recalled that one June night in 1772, *Gaspee* grounded when chasing a suspected smuggler. A Providence merchant by the name of John Brown had a local shipmaster get eight long boats ready to assault the schooner. About the "time of the shutting up of the shops ... a man passed along the main street beating a drum and informing the inhabitants of the fact that *Gaspee* [italics mine] was aground on Namquit Point ... inviting those persons who felt a disposition to go and destroy that troublesome vessel, to repair" to the rendezvous point. The armed mob, including future Continental Navy admiral Abraham Whipple, ambushed *Gaspee* by sea, wounded Dudingston, and burnt the schooner.[19]

While one might argue that this was merely an angry mob of Whiggish merchants that burnt a royal vessel, this strategy, which utilized a drummer rallying volunteers on a whim to fight off an imminent maritime threat, actually fit within the region's long history of emergency fleets. Take for example a case from 1704 when a French privateer was reported off the coast. Governor Samuel Cranston was "immediately caused the Drum to beat for Voluntiers, under the Command of Capt. [William] *Wanton*, and in 3 or four hours time Fitted and Man'd a Brigantine, with 70 brisk young men well Arm'd." Two years later, when Rhode Island came under numerous legal attacks, Cranston would cite Rhode Island's frequent "fitting and sending out vessels upon the discovery, and to secure the coast" as evidence of the colony's utility and loyalty to the Crown.[20] How ironic that Rhode Island elites would use the same strategy to resist the Crown sixty-eight years later.

For many reasons, the 1772 attack on *Gaspee* mirrored Rhode Island's emergency fleets of Queen Anne's War. On a familial level, the 1704 emergency fleet captain, William Wanton, was the father of Governor Joseph Wanton—the provincial politician who continually denied that Rhode Islanders had planned a naval assault on the British and who would deal with the immediate fallout from the *Gaspee* fiasco.[21] On a strategic level, both naval expeditions relied on a local authority having a drummer rally volunteers on a whim and piling them into boats or a vessel to fight off an immediate threat to colonial commerce. Rhode Island, like so many other colonies, had traditionally raised (or impressed) emergency fleets when Royal Navy vessels were absent in order to defend the coasts during the emergency and had now turned that same strategy on the Royal Navy itself.

For some scholars, the *Gaspee* affair fits into larger discussions of mob violence in the decade leading up to the American Revolution. In her well-known study on pre–Revolutionary War mob violence, historian Pauline Maier argues that the attack was one of many typical eighteenth-century crowd uprisings throughout the Atlantic world. These uprisings involved elites and commoners acting in concert to solve a local problem (e.g., to fight impressment) rather than to advance "revolutionary goals." Nevertheless, Maier argues, Anglo-American riots against British authorities during this era took on a new meaning as they were fights against impositions from an "external power."[22]

While the *Gaspee* riot may have demonstrated how an angry Anglo-American crowd could use traditional patterns of mob violence against imperial officers, it also demonstrated the continuity of provincial naval defense strategies throughout the eighteenth century. It should be noted here that while Anglo-American dissidents drew on historic examples of provincial naval service to protest British taxation and employed traditional naval tactics to combat Royal Navy commerce vessels, none of these activities or examples would have been possible without a large pool of common sailors willing to resist British authority. For decades, scholars have asserted that the opening moves on the path to the Revolutionary War began on the docks of colonial ports where common sailors had so long taken part in disorderly riots against authorities.[23]

Of course, there were numerous other factors that led Jack Tars to spearhead violent protests against British authority. On a macroscale, economic issues plagued northern port cities in particular during the final years of the Seven Years' War and after. Just as provincial naval and privateering expeditions declined after 1759, wartime industries in port cities that had blossomed to support the war-effort (e.g., shipbuilding) declined as Anglo-American forces conquered French Canada.[24] Historian Jesse Lemisch has argued that a perfect storm arose for maritime discontent in the mid-1760s: postwar unemployment for tens of thousands of former privateers, new British trade restrictions, and reduced shipping opportunities thanks to colonial nonimportation protests and the Stamp Act Crisis.

In New York City during the late 1760s and early 1770s, unemployed sailors engaged in violent protests and riots against the British garrison. Aside from political qualms with the red-coated garrison, sailors competed with off-duty British infantrymen for part-time jobs and resented the competition. In essence, common sailors had personal, economic, and political motives to protest British policies and taxation without representation.[25] Lemisch has argued that one of the many unifying factors for these common seamen was their shared experience of serving on privateer vessels during the Seven Years' War. While the difference between privateers and government-funded provincial navies has been maintained throughout this book, it is worth noting that the legacy of nonroyal naval forces—whether involving provincial navies or privateers—continued to shape the way Americans protested British authority throughout the imperial crisis.

Even as economic and political concerns unique to the 1760s drove many sailors to resist British authority in the streets of port cities, one even larger and traditional bogeyman continued to foster common sailors' resentment to British authority: impressment. Numerous scholars have pointed to the 1747 Knowles Riot as the prototype for maritime crowd actions against Royal Navy press gangs in later decades. By the late 1760s, violent brawls with authorities, effigy burnings, and bonfires became common tropes in sailor-initiated riots in ports from Maine to South Carolina.[26] It will be recalled that the Knowles Riot, one of the largest pre-Revolutionary riots of this sort, had its own roots in the Royal Navy's violent attempts to impress Massachusetts provincial navy sailors. Whether members of privateer crews, provincial guard ships, or merchant vessels, common sailors could find much common cause in the fight against Royal Navy conscription.

Unsurprisingly, this violent resistance to impressment would continue throughout the final years leading up to the Revolution. With the 1746 impressment act still in place, Royal Navy ships impressed hundreds of American sailors, sometimes even sending them back to Britain. Just as it had in Boston in 1747, this policy led New Yorkers to form a violent mob and burn a Royal Navy tender in the summer of 1764 and inspired similar riots throughout the colonies as far south as Virginia. Violent resistance to impressment could occur on the water, as well. As late as 1775, armed American mariners in whaleboats in Marblehead, Massachusetts, surrounded a Royal Navy vessel and rescued their impressed compatriots.[27] Even if provincial navies had largely ceased to function by the 1760s, the vital participation of provincial naval veterans in the Knowles Riots of the 1740s helped to stoke the flames of provincial anger against the Royal Navy that still burnt hot two decades later.

Ultimately, Britain's taxation policies and the Royal Navy's attempts to enforce these policies at sea failed to reduce Anglo-American dissidents to submission. While the customs authorities and the Royal Navy did make some headway in enforcing the Sugar Act, the costs of maintaining a large peacetime Royal Navy fleet were probably higher than any revenue made by subsequent imperial tax laws such as the Townshend Acts. The larger goal of connecting the American colonies to the metropole through increased imperial domination also failed as the thirteen mainland American colonies became more and more alienated.[28] Although Anglo-Americans could not convince the metropole to lighten its taxation policies by invoking their decades of provincial naval service to the Crown, they did use old provincial naval strategies and techniques to violently resist the Royal Navy's enforcement of these new imperial acts.

From Provincial Navies to the Continental and State Navies, c. 1775–1776

In 1813, aged former president John Adams reflected on the origins of the Continental Navy and remarked that "I think a circumstantial history of naval operations in this Country ought to be written even as far back as the province ship under Captain Hollowell, &c., and perhaps earlier still."[29] Adams, himself a veteran organizer of the Continental Navy in 1775, realized that the "keel" of the American Navy had been laid in the decades and even century before the Revolutionary War.

The connection between the colonies' provincial navies and the navies of the American Revolution becomes even more evident when one realizes that historians throughout the last century have typically placed coastal New England—the region with the earliest and largest provincial naval establishments in the colonies—as the birthplace of the Continental Navy. While the first shots of the war were fired at Lexington, Concord, and Bunker Hill throughout the spring and summer of 1775, New Englanders in whaleboats attacked British shipping near Boston. When the Second Continental Congress formed the Continental Army out of New England militia units and placed General George Washington as commander, the Virginia military veteran drew on the Congress's limited funds to fit out merchant ships as warships to challenge the British stranglehold around Boston.[30]

The initial fleet of New England vessels, like the provincial fleet that attacked Louisbourg in 1745, was one of merchant ships. Historian Christopher Magra has recently argued that despite their civilian origins, this merchant fleet was "the first American navy" of the war. In a method that "defies classification as privateers," patriotic merchants leased their vessels to the Continental Congress, making them "temporary property of the United Colonies." As this fleet grew, the need to clothe, feed, and pay sailors and shipwrights were some of the many factors that elevated Congress's role as a central power.[31] One might recall that in 1690, the costs associated with the New England assault on Quebec—largely a naval campaign—led the Massachusetts government to issue the first ever paper money in the colonies. Just as with provincial navies of decades past, fitting out what would become the Continental Navy would inspire lasting governmental and financial change.

While the fitting out of merchant ships as warships in Boston in the summer of 1775 would start the slow process of the formation of an American Navy, the concept of commissioning a continental fleet was just as controversial at the beginning of the American Revolution as it had been during the Albany Conference of 1754. That summer, Rhode Island's provincial government—still dealing with ravages by Royal Navy ships—urged its delegates to the Continental Congress to campaign for a Continental Navy to help deter these attacks.[32]

Some congressmen such as Pennsylvania's John Dickinson were hesitant to escalate the war effort anymore when there might still be a chance at peace through diplomacy. Firebrands such as Massachusetts's John Adams and his cousin Samuel Adams contended that creating a Continental Navy would be a show of force that would stand a better chance of achieving peace through strength. While one of Pennsylvania's delegates plied for diplomatic resolutions to the violence, one of the colony's other representatives—Benjamin Franklin—backed John Adams and other hawkish congressmen who called for a pancolonial navy to contest the Royal Navy's growing stranglehold in the Northeast. It should come as no surprise that Franklin, who had come up with the idea of a Continental Navy to support the British war effort twenty-one years before, would be more than willing to use the same idea while fighting them in 1775.[33]

By October 1775, Congress created a Naval Committee of men who represented several colonies, including South Carolina's Christopher Gadsden—a veteran of the Royal Navy himself—and Stephen Hopkins—the former governor of Rhode Island. Hopkins, then nearly seventy years old, would have been well aware of the benefits of provincial naval warfare, having served as governor while his colony hired a guard vessel during the Seven Years' War, and having attended the Albany Congress where talk of a proto–Continental Navy had occurred.[34]

While the Continental Congress worked on making a national fleet a reality, eleven of the thirteen state governments (excepting New Jersey and Delaware) took the initiative themselves to build local fleets. Whereas the Continental Congress had greater resources to build larger warships with larger crews than state fleets, local governments bankrolled large flotillas of smaller vessels (e.g., galleys) to defend regional coasts and ports.

Just as with colonial provincial navies of previous wars, these state-funded fleets were widely outnumbered by locally commissioned privateers.[35] Nevertheless, regionally focused state navies carried the legacy of provincial navies forward into the fight against the British even more than the Continental Navy.

In Massachusetts, legislators ironically justified the fitting out of privateers to raid British commerce by drawing on historical royal support for Anglo-American private naval expeditions. In the autumn of 1775, provincial congressman Elbridge Gerry—a future vice president of the United States—introduced the assembly's "An act for encouraging the fitting out of armed vessels to defend the sea coast of America" with a historical explanation for the legislation. Gerry's preamble included the following passage: "whereas [in the 1690s] their majesties, King William and Queen Mary, by the royal charter of this colony for themselves their heirs and successors, did grant ... that in the absence of the governour and lieutenant governor of the colony, a majority of the council shall have full power ... for the special defence of their said province ... to encounter, expulse, resist and pursue by force of arms, as well by sea as by land ... every such person and persons ... as should in a hostile manner invade."[36] While Gerry's foreword was for a law on privateering rather than specifically for the creation of a state navy, he likely would have used the same justification—more than a century of local naval expeditions condoned by the Crown—to remind readers that there was nothing novel about provincial naval defense.

Even though one can find echoes of provincial naval traditions in the legal origins of the Continental and state navies of the Revolution, it is equally striking that a number of surviving provincial navy veterans served in both state and national naval branches throughout the war. One of the most well-known examples is South Carolina's Captain John Joyner. His experience also exemplifies the contemporary tensions between many American officials over the importance of regional versus national maritime defense needs. Near the end of the Seven Years' War, Joyner commanded one of the colony's scout boats but saw little service other than making coastal surveys.

When the Revolutionary War began in 1775, however, a provincial government committee ordered Joyner and his compatriot John Barnwell to coordinate an assault with Georgia forces on a British supply ship near Savannah. Forces from both colonies captured over twenty thousand pounds of gunpowder from the British ship and sent at least five thousand pounds to help Washington's army then surrounding Boston. Ultimately, both colonies' local naval forces scored a local victory for both their respective governments and the American cause in general.[37]

Although Joyner's Revolutionary career began with promise, it would end in catastrophe. By 1778, South Carolina's government hoped to use valuable local staple crops (indigo and rice) and credit to purchase a few frigates from the French government. The state sent its commodore Alexander Gillon and several representatives for the transaction, including the well-experienced Joyner. On the journey over, Joyner faced a mutiny and temporary imprisonment by British authorities but was luckily able to take advantage of family connections in Bristol to secure his release.[38]

Unfortunately for Gillon and the state government, it would take three years to acquire even a single frigate. After much haggling, Gillon secured the lease for *L'Indien*—which would later be rechristened as *South Carolina*—a frigate that the French government had ordered to be constructed in neutral Amsterdam and that had been placed under the temporary guardianship of the Chevalier de Luxembourg.[39] The French had initially built the frigate with the intention of selling it to the prominent American envoy Benjamin Franklin and his compatriots, but various economic and diplomatic issues—namely Dutch neutrality until 1780—delayed this plan. It is interesting that Franklin himself had secured the initial audience with the French government for Gillon, especially since he had criticized the commodore for seeking such a large vessel for local defense purposes.[40]

By mid-1782, Commodore Gillon reached Philadelphia and there became enmeshed with the Chevalier in legal battles over some of his transactions in Europe. With these troubles in mind, the commodore placed Joyner as *South Carolina*'s captain. While Joyner aimed to bring it back to South Carolina, three Royal Navy vessels captured his vessel and its 450-man crew that December. Only a few months later, the war would be over.[41] Ultimately, Joyner's tenure as captain of *South Carolina* was brief, disastrous, and limited to the final months of the Revolutionary War. What is noteworthy, however, is that a state that had previously only built provincial navies for regional campaigns in the southeastern colonies now had the ability to secure warships from European powers and to do all this while depending on the experience of a provincial navy veteran.

South Carolina's provincial sailors were not the only veterans who saw service in the Revolution. For instance, historian Philip Chadwick Foster Smith has argued that Massachusetts's provincial navy of the Seven Years' War set a "precedent for the Massachusetts State Navy of the Revolutionary War." While Captain Hallowell of *King George* frigate stayed loyal to the king, his pilot, Eleazor Giles, became a Patriot privateer; a twelve-year-old servant onboard named Samuel Tucker later captained two of Washington's schooners and a Continental frigate; and the province ship's lieutenant, Daniel Souther, became a major captain in the state's Revolutionary navy.[42] Ironically, Souther's own warship would be a brigantine known as *Massachusetts*. It will be recalled that during the War of Jenkins' Ear, the colony's largest warship had that same name.

While the legacies and small cadre of veterans of pre-Revolutionary provincial navies shaped the development of the Continental and state navies, at least one surviving provincial navy vessel—the Georgia scout boat *Prince George*—was still in service during the War for Independence. At more than thirty years old, *Prince George* had long exceeded the lifespan of most wooden vessels of the era.[43] At the beginning of the conflict, rebels had forced its Loyalist captain, John Lichtenstein, to surrender the aged boat.[44] Shortly thereafter, *Prince George* was put to work not only guarding the Georgia coast but in assisting neighboring South Carolina's war effort. When Captain William Henry Drayton sought volunteers from Savannah to fill out the ranks of the South Carolina Navy warship *Prosper*, Georgia revolutionary Joseph Habersham reported that "I have procured the Scout Boat to go with them as far as Purisburgh" on the border between the two states.[45]

While the scout boat proved useful to both rebellious states in early 1776, disaster would strike by May of that year when Captain John Stanhope of HMS *Raven* captured the vessel during an engagement with Georgia's state navy.[46] Despite this loss, by 1778, Georgia's Whig House of Assembly ordered the state's commissary general to "make Enquiry whether the Scout Boat which before the revolution was in the Service of this State (then province) can be got up ... and also the repairs of the said Boat at the public Charge." While it is not clear if this was *Prince George* or some other scout boat, the fact that Georgia officials recognized the utility of the scout boat service to "this State (then province)" demonstrated an acknowledgement of the importance of the legacy of provincial navies years into the American Revolutionary War.[47]

With likely dozens of provincial navy veterans playing important roles in both the Continental and state navies, it might be tempting to see provincial naval service as a pipeline into support for the American cause. The case of the Braddock-Lyford family of Georgia and South Carolina challenges this simple assumption and demonstrates how family legacies of provincial naval service had little bearing on one's allegiance. While John

Braddock commanded one of the Georgia state navy galleys in 1776, his maternal uncle William Lyford Jr. was a Loyalist exile who acted as a pilot for numerous Royal Navy vessels throughout the war. Both men's fathers—David Cutler Braddock and William Lyford Sr.—had been captains in the provincial navy of South Carolina during the 1740s.[48] Even though their fathers fought in the same fleet against the Spanish, differing political loyalties in the 1770s would tragically drive these close relatives to join opposing navies during the Revolution.

The legacy of provincial naval service was also evident beyond the American Whigs' cause. For example, the governor of Loyalist-aligned East Florida, Patrick Tonyn, faced a traditional dilemma between 1776 and 1778 when Royal Navy vessels either failed to protect his colony's coast or their vessels were in ill shape to assist him. To defend St. Augustine and the rest of Britain's only loyal colony on the North American mainland south of Nova Scotia, Tonyn created a fleet of privateers, impressed vessels, and purchased warships that some historians have called the "East Florida provincial navy." Despite the absence of the Royal Navy, the Loyalist provincial navy of East Florida successfully repelled numerous rebel American invasions until the British secured the province by capturing Savannah, Georgia, in 1778. In the end, Tonyn's provincial navy was no different from historical provincial navies in the now-rebellious colonies or his opponents' state navies.[49]

For a long period between 1689 and 1754, Anglo-Americans fitted out their own semipermanent and temporary provincial navies to secure their coasts from French, Spanish, piratical, and Native American maritime threats with limited royal assistance. During the Seven Years' War (c. 1754–63), the Royal Navy finally gained maritime hegemony in the western Atlantic world and made the existence of costly colonial naval establishments unnecessary. Nevertheless, when the British government used the Royal Navy to enforce unpopular trade policies in the 1760s, Anglo-American antipathy for the navy's heavy-handed impressment policies and enforcement of trade laws coupled with a long legacy of local naval defense shaped the patterns of Anglo-American naval resistance during the imperial crisis. When the imperial crisis became a War for Independence, American partisans—ranging from Founding Fathers to provincial naval veterans—drew on regional and personal provincial naval experience to organize the Continental and state navies of the Revolutionary War.

Conclusion

A century ago, historian Howard Chapin remarked that "the American Navy did not spring forth full-fledged at the outbreak of the Revolution, like Pallas Athene from the head of Zeus. Its roots go back to the Colonial privateersmen and the naval expeditions against the French and Spanish."[1] Chapin was mostly correct, except for the fact that the concept of local naval defense had even longer roots in the regional and private fleets of medieval England. Throughout the seventeenth century, English colonists throughout the Atlantic world continued those traditions of local naval defense in their fleets of small craft that hunted pirates and transported soldiers in early conflicts with Indigenous people.

By the end of the seventeenth century, the rise of global imperial conflicts such as King William's War and Queen Anne's War would force Anglo-American governments from Canada to the West Indies to expand upon these defensive fleets—using limited local funds and emergency powers to impress vast emergency fleets and even to construct warships to protect their coastlines. During these global wars, provincial governments also orchestrated massive expeditions against French and Spanish strongholds such as Quebec and St. Augustine. Whether done independently or with the help of small Royal Navy squadrons, these missions were rarely successful and often times exacerbated sociopolitical and economic tensions within the colonies themselves.

Even though there was a lull in imperial warfare in the interwar period between 1713 and 1739, unresolved tensions on imperial borderlands in Canada, New England, South Carolina, Georgia, and the West Indies ensured continued violence between Anglo-Americans and various imperial, piratical, and Indigenous enemies. Limited royal assistance during this crucial period meant that provincial governors, councils, and legislatures would be forced to continue to foot the bill for provincial naval forces and to adapt them to meet those immediate threats.

Renewed imperial warfare against Spain in 1739 and France in 1744 brought with it a massive rise in the scale of provincial naval warfare. By the War of Jenkins' Ear, London did make some strides to expand the Royal Navy's presence in North America and to support the efforts of *some* colonial defense fleets. Nevertheless, violent disagreements over strategy and impressment between provincial and Royal Navy officials coupled with legal vagueness over the status of provincial navies limited the effectiveness of any naval cooperation between the metropole and the colonies. Even when the Royal Navy spearheaded the victorious naval campaigns of the final colonial war—the Seven Years' War—Anglo-American resentment over impressment and jealousy for provincial autonomy created a distaste for the Crown's naval forces.

After the Seven Years' War, Parliament passed a series of taxes on Anglo-Americans to cover the costs of the Seven Years' War and various other unpopular imperial legal reforms. When London used the Royal Navy to enforce these unpopular laws, decades of resentment towards Crown forces led Whig dissidents to violently protest these measures at sea throughout the imperial crisis of the 1760s and 1770s. During these protests and the early years of the War for Independence, American acts of resistance to British Navy—ranging from violent protests to creating Continental and state navies—all drew on traditions, laws, and even the experience of veterans from America's provincial naval past.

Throughout this book, I have told the story of how Anglo-American colonists depended on their own naval resources to maintain coastal security and to fight local and imperial enemies. It is not my contention that these pre-Revolutionary forces were direct progenitors of the Continental (and future US) Navies. Nevertheless, I do believe that more than a century and a half of mostly autonomous regional naval defense provided a necessary experiential foundation for the Founders to draw on when creating the first national and state fleets of the Revolutionary Era.

On its own, the story of provincial navies is unique, and a study of pre-Revolutionary American colonies creating naval institutions is novel. Nevertheless, as I have tried to demonstrate throughout *The First Fleets*, the story of provincial navies can also give us new maritime avenues with which we can reexamine the important social, political, racial, and religious dynamics at play in the pre-Revolutionary Atlantic world. For instance, some of the first paper money issued in North America was printed in New England to cover the costs of the failed naval expedition against Quebec in 1690. In another example, South Carolina's scout boat navy was used to pursue African escapees from chattel slavery as much as it was used to combat Spanish incursions on Britain's contested southern borderlands. These examples demonstrate how these little-studied naval organizations were intimately connected with the major sociopolitical issues of the societies they emerged from.

Ironically, most of my research and composition of the book have taken place during the preliminary years leading up to the semiquincentennial (250th anniversary) of the American Revolution. Historians, legislators, living historians, and educators have all started to discuss the best ways to remember this pivotal moment in our national history. It is my hope that during these commemorations—and well beyond—historians will recognize that the American naval tradition did not begin with the rebel fleets of the Revolutionary War but much further back in the annals of colonial history.

Notes

GLOSSARY OF NAVAL TERMINOLOGY

1. William Avery Baker, "Vessel Types of Colonial Massachusetts," in *Collections of the Colonial Society of Massachusetts*, vol. 52, *Sea Faring in Colonial Massachusetts* (March 1980), 18–20; Colonial Society of Massachusetts website. John Robinson and George Francis Dow, *The Sailing Ships of New England, 1607–1907* (New York: Skyhorse Publishing, 2007), 28–29.

2. David Wilson "Protecting Trade by Suppressing Pirates: British Colonial and Metropolitan Responses to Atlantic Piracy, 1716–1726," in *The Golden Age of Piracy: The Rise, Fall, and Enduring Popularity of Pirates*, ed. David Head (Athens: University of Georgia Press, 2018), 91, Google Play eBook edition.

3. Baker, "Vessel Types," 22; and Benerson Little, *Pirate Hunting: The Fight against Pirates, Privateers, and Sea Raiders from Antiquity to the Present* (Washington, DC: Potomac Books, 2010), 147, Google Play eBook edition.

4. P. C. Coker, *Charleston's Maritime Heritage, 1670–1865: An Illustrated History* (Coker Craft, 1987), xii–xiv; and Baker, "Vessel Types," 12–13.

5. Little, *Pirate Hunting*, 140–41; and Larry Ivers, *This Torrent of Indians: War on the Southern Frontier, 1715–1728* (Columbia: University of South Carolina Press, 2016), 104–5, Kindle eBook edition.

6. Benerson Little, *The Buccaneer's Realm: Pirate Life on the Spanish Main, 1674–1688* (Washington, DC: Potomac Books, 2007), 15.

7. Baker, "Vessel Types," 10–11.

8. Coker, *Charleston's Maritime Heritage*, xii–xiv.

9. Baker, "Vessel Types," 13–15.

10. Baker, "Vessel Types," 18–19; Coker, *Charleston's Maritime Heritage*, xii–xiv. Confusingly, the Royal Navy also used the term sloop-of-war to describe a wide array of small warships in this era. See Ian McLaughlan, *The Sloop of War, 1650–1763* (Barnsley: Seaforth Publishing, 2014), for more information on Royal Navy sloops.

11. Waldo Lincoln, *The Province Snow "Prince of Orange"* (Worcester: Press of Charles Hamilton, 1901), 4.

12. Ivers, *This Torrent*, 106–7.

INTRODUCTION

1. Wilson, "Protecting Trade," 98–99.

2. Carl E. Swanson, *Predators and Prizes: American Privateering and Imperial Warfare, 1739–1748* (Columbia: University of South Carolina Press, 1990), 160–61.

3. Charles O. Paullin, *Colonial Army and Navy*, unpublished manuscript, Charles Oscar Paullin papers, 1931, MSS53033, Library of Congress, 45.

4. N. A. M. Rodger, *The Command of the Ocean: A Naval History of Britain, 1649–1815* (New York: W. W. Norton, 2004), 232.

5. Benjamin Franklin, "Plain Truth, 17 November 1747," Founders Online, National Archives website.

6. John R. Dull, *American Naval History, 1607–1865: Overcoming the Colonial Legacy* (Lincoln: University of Nebraska Press, 2012), 2–10. Although I challenge Dull's statement on pre-Revolutionary naval forces, I recognize that this was a very minor part of his much larger monograph that contextualizes the limits of early American naval strategies throughout the early national period.

7. N. A. M. Rodger, "The New Atlantic: Naval Warfare in the Sixteenth Century," in *War at Sea in the Middle Ages and the Renaissance*, ed. John B. Hattendorf and Richard W. Unger, 238–47 (London: Boydell & Brewer, 2003), JSTOR.

8. Rodger, "The Law and Language of Private Naval Warfare," *Mariner's Mirror* 100, no. 1 (2014): 5–13. More recently, John Coakley has made the case that the term *privateer* emerged in the English lexicon because of private naval warfare in Jamaica in the 1660s. See Coakley, "'The Piracies of Some Little Privateers': Language, Law and Maritime Violence in the Seventeenth-Century Caribbean," *Britain and the World* 13, no. 1 (2020): 6–26, EBSCOhost. Also see Shinsuke Satsuma, *Britain and Colonial Maritime War in the Early Eighteenth Century: Silver, Seapower and the Atlantic* (London: Boydell & Brewer, 2013), 9–10, JSTOR.

9. Rodger, "Law and Language," 5.

10. Edward Phillips, *The New World of Words: Or Universal English Dictionary* (London: King's Arms, 1720), no page number given, Google Books eBook. Samuel Johnson, *A Dictionary of the English Language: A Digital Edition of the 1755 Classic* (1755; repr., Johnson Dictionary Online, 2012), 1573.

11. Howard Chapin, *Privateer Ships and Sailors: The First Century of American Colonial Privateering, 1625–1725* (1926; repr., Martino Fine Books, 2017), 7–8.

12. Chapin, *Privateer Ships*, p. 96.

13. Paullin, "Colonial Army and Navy," 71. This also contrasts with Rodger's argument that the Royal Navy rarely even patrolled American waters before 1713. See Rodger, *Command of the Ocean*, 232.

14. W. A. B. Douglas, "The Sea Militia of Nova Scotia, 1749–1755: A Comment on Naval Policy," *Canadian Historian Review* 47, no. 1 (March 1966): 22–37; Larry E. Ivers, *British Drums on the Southern Frontier: The Military Colonization of Georgia, 1733–1749* (Chapel Hill: University of North Carolina Press, 1974), 165; and Swanson, *Predators and Prizes*, 50.

15. Admittedly, even the term *provincial navy* has its limitations. While one could

make the case that colonial transport and supply vessels should be included in this definition, I have decided to limit its scope in this definition to vessels that colonial governments specifically designated for combat missions. Additionally, there were times when colonial governments hired or impressed *privateer* ships into direct state service during emergencies. During situations in which independent privateers came under direct government control or command, I consider these vessels and their crews to be part of provincial navies.

16. In a 2012 essay on the origins of the Continental Navy, historian John Hattendorf noted that there had been "no comparable development for a theoretical understanding of the roles of small navies," and used the small American fleet of the Revolutionary War as a case study of state formation and naval warfare in the early modern period. This book similarly looks at the ways in which pre-Revolutionary Anglo-American naval organizations challenged and contributed to provincial state building. See John Hattendorf, *Talking About Naval History: A Collection of Essays* (Pittsburgh: US Government Printing Office, 2012), 188–93.

17. Historians have long debated the ability of the British Crown to impose its military and legislative will on colonial America. While a few scholars such as Stephen Saunders Webb have argued that the Crown had significant military and political control over its American provinces from the seventeenth century onward, most other scholars contest this view and echo Jack Greene's view that power in the British Atlantic world was negotiated between "periphery and center." For examples of this debate see Stephen Saunders Webb, *The Governors-General: The English Army and the Definition of Empire, 1569–1681* (Chapel Hill: University of North Carolina Press, 1979); Jack M. Sosin, *English America and the Restoration Monarchy of Charles II: Transatlantic Politics, Commerce, and Kinship* (Lincoln: University of Nebraska Press, 1980); Jack Greene, *Peripheries and Center: Constitutional Development in the Extended Polities of the British Empire and the United States, 1607–1788* (New York: Norton, 1990). Kurt Nagel's study, *Empire and Interest: British Colonial Defense Policy, 1689–1748* (PhD diss., Johns Hopkins University, 1992), was among the earliest to expand upon this debate in terms of military policies and has been extremely influential in my research.

18. Historians have long discussed the role of the British Royal Navy in shaping Britain's policies in North America (particularly after the Seven Years' War) but have not yet considered the role of Anglo-American fleets in this same arena. For examples, see Neil R. Stout, *The Royal Navy in America, 1760–1775* (Annapolis: Naval Institute Press, 1973); Daniel Baugh, "Great Britain's 'Blue-Water' Policy, 1689–1815," *International History Review* 10, no. 1 (1988): 33–35, JSTOR; Eliga Gould, *The Persistence of Empire: British Political Culture in the Age of the American Revolution* (Chapel Hill: University of North Carolina Press, 2000); and Sarah Kinkel, *Disciplining the Empire : Politics, Governance, and the Rise of the British Navy* (Cambridge: Harvard University Press, 2018). Over the last two decades, historians have also widely expanded their consideration of the role of maritime conflict (and predation) in shaping the political and social landscape of the British Atlantic world. These studies have taken multiple forms, from scholarly examinations of the role of maritime piracy and its suppression to groundbreaking reinterpretations of Indigenous maritime power on contested borderlands.

For examples of this "maritime wave," see Mark Hanna, *Pirate Nests and the Rise of the British Empire, 1570–1740* (Chapel Hill: University of North Carolina Press, 2015); Andrew Lipman, *The Saltwater Frontier: Indians and the Contest for the American Coast* (New Haven: Yale University Press, 2015); and Matthew Bahar, *Storm of the Sea: Indians and Empires in the Atlantic's Age of Sail* (Oxford, Oxford University Press, 2019).

Chapter 1

1. John Winthrop, *Winthrop's Journal, History of New England, 1630–1649*, vol. 1, ed. James Kendall Hosmer (New York: Charles Scribner's Sons, 1908), 95. Chapin, *Privateer Ships*, 17–18.

2. George Francis Dow and John Henry Edmonds, *Pirates of the New England Coast, 1630–1730* (Salem: Marine Research Society, 1923), 22.

3. See Richard Harding, *The Evolution of the Sailing Navy, 1509–1815* (New York: St. Martin's Press, 1995), 1–4; Susan Rose, *England's Medieval Navy, 1066–1509: Ships, Men & Warfare* (Montreal: McGill-Queen's University Press, 2013), 32–43; Douglas Burgess Jr., *The Politics of Piracy: Crime and Civil Disobedience in Colonial America* (Lebanon: University Press of New England, 2014), 17–19, ProQuest.

4. Rodger, "New Atlantic," 237.

5. Harding, *Evolution of the Sailing Navy*, 1–4; Rodger, "New Atlantic," 37.

6. Rodger, *The Safeguard of the Sea: A Naval History of Britain, 640–1649* (New York: W. W. Norton, 1997), 19–27.

7. Rodger, *Safeguard of the Sea*, 117–25.

8. Harding, *Evolution*, 4–5; Rose, *England's Medieval Navy*, 32–34.

9. Rose, *England's Medieval Navy*, 179–82.

10. Jill Eddison, *Medieval Pirates: Pirates, Raiders and Privateers, 1204–1453* (Stroud: History Press, 2013), 27–35, Kindle eBook edition.

11. Burgess, *Politics of Piracy*, 14–18, ProQuest.

12. David J. Starkey "Voluntaries and Sea Robbers: A Review of the Academic Literature on Privateering, Corsairing, Buccaneering and Piracy," *Mariner's Mirror* 97, no. 1 (2011): 127–29.

13. Burgess, *Politics of Piracy*, 20.

14. Rose, *England's Medieval Navy*, 142–43.

15. Harding, *Evolution*, 6–7.

16. Phyllis Jestice, "Naval Warfare," in *Fighting Techniques of the Medieval World, AD 500–AD 1500*, ed. Matthew Bennett et al. (New York: St. Martin's Press, 2005), 247–50.

17. Jan Glete, *Warfare at Sea, 1500–1650: Maritime Conflicts and the Transformation of Europe* (London: Taylor & Francis Group, 1999), 132–35, ProQuest. There has been a robust historiographical debate about the nature of state development and warfare during the early Modern period—particularly around the controversial military revolution theory that posits that a sudden evolution in military technology and tactics between 1560 and 1660 spurred on massive societal transformations in Europe. For a good overview of the naval element of this historiographical debate, see Eugene Rasor, *The Seaforth Bibliography: A Guide to More Than 4000 Works on British Naval History 55BC—1815* (Pen & Sword Books Limited, 2008), 261–63.

18. Harding, *Evolution*, 11–13.

19. Rodger, *Safeguard of the Sea*, 222–28.

20. Hanna, *Pirate Nests*, 36; and Rodger, "Law and Language," 9.

21. Harding, *Evolution*, 20–22; and Michael J. Braddick, *State Formation in Early Modern England, c. 1550–1700* (Cambridge: Cambridge University Press, 2000), 205–9, ProQuest.

22. Hanna, *Pirate Nests*, 36–38.

23. Hanna, *Pirate Nests*, 45–46; and John D. Grainger, *The British Navy in the Caribbean* (Woodbridge: Boydell and Brewer, 2021), 22–42, Kindle eBook edition.

24. Glete, *Warfare at Sea*, 258–61; and James McDermott, *England & the Spanish Armada: The Necessary Quarrel* (New Haven: Yale University Press, 2005), 198–201.

25. Kenneth R. Andrews, *Ships, Money, and Politics: Seafaring and Naval Enterprise in the Reign of Charles I* (Cambridge: Cambridge University Press, 1991), 143–44; Harding, *Evolution*, 32–40; Braddick, *State Formation*, 206–10; and Hanna; *Pirate Nests*, 69–70.

26. Braddick, *State Formation*, 207–8; Andrews, *Ship, Money, and Politics*, 144–45; and Harding, *Evolution*, 38–40.

27. Harding, *Evolution*, 45–49.

28. Andrews, *Ships, Money, and Politics*, 6–11; Braddick, *State Formation*, 211–12; and Harding, *Evolution*, 51–3.

29. Hanna, *Pirate Nests*, 67–90.

30. Harding, *Evolution*, 59–61.

31. Grainger, *British Navy*, 46, 75–99; and Harding, *Evolution*, 73–83.

32. J. D. Davies, *Pepys's Navy: Ships, Men & Warfare, 1649–1689* (Barnsley: Seaforth Publishing, 2008), 611–13, 625–27, Kindle eBook edition; and Grainger, *British Navy*, 119–20.

33. Carla Gardina Pestana, *The English Atlantic in an Age of Revolution, 1640–1661* (Cambridge: Harvard University Press, 2007), 238–39, ProQuest.

34. Jack Greene, *The Constitutional Origins of the American Revolution. New Histories of American Law* (Cambridge: Cambridge University Press, 2011), 8–24, ProQuest. For more on the debate about the strength of the royal prerogative in colonial America, see Webb, *Governors-General*; Sosin, *English America and the Restoration Monarchy*; Greene, *Peripheries and Center*; Nagel, *Empire and Interest*; and Elizabeth Mancke, "Negotiating an Empire: Britain and Its Overseas Peripheries, c. 1550–1780," in *Negotiated Empires: Centers and Peripheries in the Americas, 1500–1820*, ed. Christine Daniels and Michael V. Kennedy (London: Taylor & Francis Group, 2002), 232–65, ProQuest.

35. Paullin *Colonial Army and Navy*, 66; and "The Charter of Massachusetts Bay: 1629," Yale University, Avalon Project website.

36. Kyle F. Zelner, *A Rabble in Arms: Massachusetts Towns and Militiamen during King Philip's War* (New York: New York University Press, 2009), 19–29, ProQuest.

37. *The Charters and General Laws of the Colony and Province of Massachusetts Bay* (Boston: T. B. Wait, 1814), 87.

38. Lipman, *Saltwater Frontier*, 4–5.

39. Lipman, *Saltwater Frontier*, 131–35.

40. Chapin, *Privateer Ships*, 18.

41. Lipman, *Saltwater Frontier*, 135–37.

42. Edwin Monroe Bacon, *The Connecticut River and the Valley of the Connecticut Three Hundred and Fifty Miles from Mountain to Sea; Historical and Descriptive* (New York: G. P. Putnam's Sons, 1906), 49; and David M. Powers, "'Use Dilatory Means': William Pynchon and the Native Americans," Conference Paper Presented at 17th Century Warfare, Diplomacy & Society In the American Northeast Conference, Battlefields of the Pequot War Project website.

43. Legislature Minutes, Massachusetts, May 14, 1645, in *Records of the Governor and Company of the Massachusetts Bay in New England*, vol. 3, 1644–1657, ed. Nathaniel B. Shurtleff (Boston: William White, 1854), 9–10.

44. Pestana, *English Atlantic*, 33. For more on Ninigret and his rivalry with the New England Confederation, see Julie A. Fisher and David J. Silverman, *Ninigret, Sachem of the Niantics and Narragansetts: Diplomacy, War, and the Balance of Power in Seventeenth-Century New England and Indian Country* (Ithaca: Cornell University Press, 2020).

45. *Records of the Colony of New Plymouth in New England*, vol. 2, Acts of the Commissioners of the United Colonies of New England, 1653–1679, ed. David Pulsifer (Boston: William White, 1859), 150–55. Chapin, *Privateer Ships*, 28.

46. Alan Taylor, *American Colonies: The Settling of North America* (New York: Penguin, 2001), 136, Kindle eBook edition.

47. Arthur Pierce Middleton, *Tobacco Coast: A Maritime History of Chesapeake Bay in the Colonial Era* (Newport News: Mariners' Museum, 1953), 305–12.

48. Middleton, *Tobacco Coast*, 305–7; and Thomas Ludwell to Lord Arlington, June 24, 1667, in the article "Attacks by the Dutch on the Virginia Fleet in Hampton Roads in 1667," *Virginia Magazine of History and Biography* 4, no. 3 (January 1897): 230–36. This article is a compendium of various letters describing the attack. I have not seen many cases of the king's Broad Arrow being used for ship impressment. For more on the meaning of the symbol see Arthur Charles Fox-Davies, *A Complete Guide to Heraldry* (London: T. C. & E. C. Jack, 1909), 457.

49. Minutes of the Council of Barbados, 26–27 March 1667, in *Calendar of State Papers Colonial, America and West Indies [hereafter referred to as CSP]*, vol. 5, 1661–1668, ed. W Noel Sainsbury, 451–59 (London: Her Majesty's Stationery Office, 1880), British History Online.

50. Hanna, *Pirate Nests*, 104–8.

51. Hanna, *Pirate Nests*, 107–9; and Grainger, *British Navy*, 99–101.

52. Rodger, *Command of the Ocean*, 77–79; and Kinkel. *Disciplining the Empire*, 40.

Chapter 2

1. Richard S. Dunn, "The Glorious Revolution and America," in *Origins of Empire: British Overseas Enterprise to the Close of the Seventeenth Century*, ed. Nicholas Canny (Oxford: Oxford University Press, 2001), 445–47, ProQuest.

2. Robert K. Wright Jr., *Continental Army* (Washington, DC: Center of Military History, 1983), 6, US Army, Center of Military History website. Also see Douglas Leach, *Arms for Empire: A Military History of the British Colonies in North America, 1607–1763* (New York: Macmillan, 1973), 1–41, Google Play eBook.

3. Don Higginbotham, "The Early American Way of War: Reconnaissance and Appraisal," *William and Mary Quarterly* 44, no. 2 (April 1987): 253, JSTOR.

4. William R. Miles, *The Royal Navy and Northeastern North America, 1689–1713*, unpublished master's thesis, Saint Mary's University, Halifax, Nova Scotia, 2000, 7–8, Collections Canada website.

5. Ruth Bourne, *Queen Anne's Navy in the West Indies* (New Haven: University of New Haven, 1939), 56–59; G. S. Graham, "The Naval Defence of British North America, 1739–1763," *Transactions of the Royal Historical Society* 30 (1948): 95–96, JSTOR; Peter T. Bradley, *British Maritime Enterprise in the New World from the Late Fifteenth to the Mid-Eighteenth Century* (Lampeter: Edwin Mellen Press, 1999), 197–202; and Miles, *Royal Navy*, 66.

6. Christian Buchet, "The Royal Navy and the Caribbean, 1689–1763," *Mariner's Mirror* 80, no. 1 (1994): 37. McLaughlan, *Sloop of War*, 79; Rodger, *Command of the Ocean*, 163–65; Richard Harding, *Seapower and Naval Warfare, 1650–1830* (Routledge, 1999), 164–68, ProQuest; and Satsuma, *Britain and Colonial Maritime War*, 244–47.

7. Mary Lou Lustig, *The Imperial Executive in America: Sir Edmund Andros, 1637–1714* (Madison: Fairleigh Dickinson University Press, 2002), 134–39.

8. Charles McLean Andrews, ed., "Introduction," in *Original Narratives of Early American History: Narratives of the Insurrections 1675–1690* (New York: Charles Scribner's Sons, 1915), 213–14.

9. Hanna, *Pirate Nests*, 144–48.

10. Hanna, *Pirate Nests*, 157–58, 170–71, 179–81.

11. Lustig, *Imperial Executive*, 49, 167–68.

12. Council Minutes, Dominion of New England, May 25, 1687, in "Proceedings of the Council of the Dominion of New England from 4th May to 28th July 1687," Minutes, the National Archives (hereafter TNA), Kew, CO 5/785, 62–63. Adam Matthew, Colonial America Digital Database. For further TNA entries, please note that page numbers given after CO 5 entries are primarily stamped folio numbers or internal page numbers if stamped folio numbers are not provided. These are not always readily available and are included when/where possible.

13. Council Minutes, Dominion of New England, July 28, 1687, in "Proceedings of the Council of the Dominion of New England from 4th May to 28th July 1687," TNA CO 5/785, 65-67. This vessel was evidently later reclassified as a sloop. For a succinct description of differences between coastal trading vessels see Coker, *Charleston's Maritime Heritage*, xii–xiv.

14. Edmund Andros to John Cooke, August 6, 1687, Massachusetts State Archives, vol. 127, p. 420, FamilySearch website; and "Orders for John Cooke . . .," Massachusetts State Archives, vol. 127, p. 266, FamilySearch website.

15. Lustig, *Imperial Executive*, 174.

16. I have chosen to use the modern standard of italicizing vessel names even when the original sources do not in order to prevent confusion with the names of officers or other individuals. "Sir Edmund Andros' account of the force raised in the year 1688 for the defence of New England against the Indians," Report, Military Document, TNA, Kew, CO 5/855, 244–45. Adam Matthew, Colonial America Digital Database.

154   NOTES TO CHAPTER 2

17. For more information on the sailing rigs and utilization of such coastal vessels, see Coker, *Charleston's Maritime Heritage*, xii–xiv.

18. Lustig, *Imperial Executive*, 174–79.

19. Dunn, "Glorious Revolution," 452, 455–56.

20. Richard R. Johnson, *Adjustment to Empire: The New England Colonies, 1675–1715* (Rutgers: Rutgers University Press, 1981), 90–92; Miles, *Royal Navy*, 107–8.

21. "Sir Edmund Andros' account . . ."; "The Humble Peticon and Request of John Cooke," c. summer 1689, Massachusetts State Archives Collection, Colonial Period, 1622–1788, vol. 107, Revolution, 1689–1700, 79, FamilySearch website; and Lords of Trade and Plantations to King William III, June 12, 1690, in "New England patents and grants, 1690," Submission, Order, TNA, Kew, CO 5/905, 116–18. Adam Matthew, Colonial America Digital Database.

22. Elisha Cooke and Thomas Oakes, "An Answer to Sr: Edmond Andros's Acco: of the Forces Raised in New England for Defence of the Country against the Indians . . .," May 30, 1690, 113–15 in "New England patents and grants, 1690."

23. Harriet Silvester Tapley, *The Province Galley of Massachusetts Bay, 1694–1716: A Chapter of Early American Naval History* (Salem: Essex Institute, 1922), 1–2; and Chapin, *Privateer Ships*, 108–9.

24. "Byfield's letter to Dudley about assemblies," June 12, 1694, Correspondence, TNA, Kew, CO 5/858, 102. Adam Matthew, Colonial America Digital Database.

25. Lieutenant Governor William Stoughton to Council of Trade and Plantations, September 30, 1697, in *CSP*, vol. 15, *May, 1696–31 October 1697*, ed. J. W. Fortescue (London: Mackie, 1904), 624–25, British History Online.

26. Bourne, *Queen Anne's Navy*, 32–33, 68–70.

27. For an example of a scholar who notes the importance of provincial naval vessels in the West Indies, see Norton H. Moses "The British Navy and the Caribbean, 1689–1697," *Mariner's Mirror* 52, no. 1 (1966): 13–40.

28. Rodger, *Command of the Ocean*, 160–61.

29. Governor Christopher Codrington to Lords of Trade and Plantations, November 1689, TNA, Kew, CO 153/4, 188–201.

30. Governor Parke's Reply to the 22 Articles of Complaint, June 26, 1709, in *CSP*, vol. 24, *1708–1709*, ed. Cecil Headlam, 370–408 (London: His Majesty's Stationery Office, 1922), British History Online.

31. Minutes of Council in Assembly of Barbados, September 8, 1702, in *CSP*, vol. 20, *1702*, ed. Cecil Headlam, 581–88 (London: His Majesty's Stationery Office, 1912), British History Online.

32. Unnamed editor, notes to chapter 112, *The Acts and Resolves, Public and Private, of the Province of the Massachusetts Bay*, vol. 8, *Being Volume III of the Appendix, Containing Resolves, Etc. 1703–1707* (Boston: Wright & Potter Printing Co., 1895), 337–39.

33. Benjamin Church, *The History of King Philip's War; also of expeditions against the French and Indians in the eastern parts of New-England, in the years 1689, 1690, 1692, 1696 and 1704 . . .* (repr., Boston: Howe & Norton, 1825), 235.

34. Council Minutes, Massachusetts, May 2, 1701, in "Minutes of the council of the Massachusetts Bay from 9th January to 13th May 1701," Minutes, TNA, Kew, CO 5/788, 24–25, Adam Matthew, Colonial America Digital Database.

35. *Boston News-Letter*, June 4, 1705, 2, Readex, America's Historical Newspapers.

36. Chapin, *Privateer Ships*, 148. Chapin notes that one of these sloops, *Anne*, served with Church's expedition. For more information on Quelch's background, see Hanna, *Pirate Nests*, 334–36.

37. Joseph Dudley to the Council of Trade and Plantations, July 13, 1704, in *CSP*, vol. 22, *1704–1705*, ed. Cecil Headlam, 211–23 (London: His Majesty's Stationery Office, 1916), British History Online.

38. Owen Stanwood, *The Empire Reformed: English America in the Age of the Glorious Revolution* (Philadelphia: University of Pennsylvania Press, 2013), 157–59, ProQuest.

39. Miles, *Royal Navy*, 49–50.

40. Nagel, *Empire and Interest*, 25–30.

41. "Extract of a [Letter] to Mr. Elisha Cook," in "Extracts and abstracts of letters from Boston," Correspondence, TNA, Kew, CO 5/855, Adam Matthew, Colonial America Digital Database.

42. *Publick Occurrences* (Boston), September 25, 1690, 2, Readex, America's Historical Newspapers.

43. "Massachusetts Documents, 1689–1692," ed. Robert Moody, 252–55, Colonial Society of Massachusetts website. For information on the loan and on the number of soldiers and sailors, see Stanwood, *Empire Reformed*, 164.

44. Miles, *Royal Navy*, 49–50.

45. For more problems with this paper currency, see Johnson *Adjustment to Empire*, 197–99.

46. "News from New England," February 2, 1690/1, in "Extracts and news from New [England]," Correspondence, TNA, Kew, CO 5/856, 27, Adam Matthew, Colonial America Digital Database.

47. Daniel Vickers, "The Northern Colonies: Economy and Society, 1600–1775," in *The Cambridge Economic History of the United States*, vol. I (Cambridge: Cambridge University Press, 1996), 243–44. Also see Dror Goldberg, "The Massachusetts Paper Money of 1690," *Journal of Economic History* 69, no. 4 (2009): 1092–1106, for how this strategy impacted future international usage of paper currency.

48. Charles W. Arnade, *The Siege of St. Augustine in 1702* (Gainesville: University of Florida Press, 1959), 29–44; Lt.-Governor Handasyd to the Earl of Nottingham, December 10, 1702, Enclosing Minutes of Council of Jamaica, Dec. 4 and Dec. 7, and Robert Quary to the Council of Trade and Plantations, December 7, 1702, in *CSP*, vol. 21, *1702–1703*, ed. Cecil Headlam, 13–44 (London: His Majesty's Stationery Office, 1913), British History Online.

49. Miles, *Royal Navy*, 56–57. Also see Vetch's 1708 pamphlet "Canada Survey'd [. . .]," July 27, 1708, in *CSP*, vol. 24, *1708–1709*, ed. Cecil Headlam, 40–56 (London: His Majesty's Stationery Office, 1922), British History Online, in which he proposes a joint Royal Navy expedition with provincial transport ships to take Quebec.

50. James D. Alsop, "Samuel Vetch's 'Canada Survey'd': The Formation of a Colonial Strategy, 1706–1710." *Acadiensis* 12, no. 1 (1982): 45–46, JSTOR. Also, see Adam Lyons, *The 1711 Expedition to Quebec: Politics and the Limitations of British Global Strategy* (New York: Bloomsbury Academic, 2013), 27, Google Play eBook.

51. "Instructions to Coll. Vetch to be observed in his Negociations with the Gov'r of

the several colonies dated March 1st. 1708/9," Order, TNA, Kew, CO 5/9, 87–92, Adam Matthew, Colonial America Digital Database.

52. Chapin, *Privateer Ships*, 83.

53. Satsuma, *Britain and Colonial Maritime War*, 118.

54. Col. Samuel Vetch to [the Earl of Sunderland?], August 2 and 12, 1709, in *CSP*, vol. 24, *1708–1709*, ed. Cecil Headlam, 437–57, British History Online.

55. Lyons, *1711 Expedition*, 28.

56. Miles, *Royal Navy*, 57–58.

57. Council Minutes, Massachusetts, July 19, 1710, in "Minutes of council of the Massachusetts Bay, Jul-Nov 1710," Minutes, TNA, Kew, CO 5/791, 79–80, Adam Matthew, Colonial America Digital Database.

58. Council Minutes, Massachusetts, August 7, 1710, 81–82.

59. David Marley, *Wars of the Americas: A Chronology of Armed Conflict in the New World, 1492 to the Present* (Santa Barbara: ABC-CLIO, 1998), 233–34.

60. Lyons, *1711 Expedition*, 133–34. See the rest of Lyons's book for more information on strategic and operational failures during the expedition.

61. Lyons, *1711 Expedition*, 151; and Douglas Edward Leach, *Roots of Conflict: British Armed Forces and Colonial Americans, 1677–1763* (Chapel Hill: University of North Carolina Press, 1986), 46–53, Kindle eBook edition.

62. Nagel, *Empire and Interest*, 286; and Lyons, *1711 Expedition*, 156–57. For more information on the funding dispute with the *Province Galley*, see Sinclair Hitchings, "Guarding the New England Coast: The Naval Career of Cyprian Southack," in *Publications of the Colonial Society of Massachusetts*, vol. 52, *Seafaring in Colonial Massachusetts, A Conference Held by the Colonial Society of Massachusetts November 21 and 22, 1975* (Boston: Colonial Society of Massachusetts, 1980), 43–59, Colonial Society of Massachusetts website.

63. See Theodore Jabbs, *South Carolina Colonial Militia, 1663–1733*, PhD diss., University of North Carolina, 1973, 260–62. Additionally, Kenneth R. Jones discusses the three major sources that described this battle in his article "A 'Full and Particular Account' of the Assault on Charleston in 1706," *South Carolina Historical Magazine* 83, no. 1 (January 1982): 1–11, JSTOR. Jones includes a transcription of the primary source "A Full and Particular Account of an Invasion Made by the French and Spaniards upon South Carolina, with the Disappointment and Disgrace they Met With in it, Contain'd in a Letter from Charles Town, in Carolina, September 12, 1706."

64. Walter B. Edgar, *South Carolina: A History* (Columbia, S.C.: University of South Carolina Press, 1998), 39–40. Taylor, *American Colonies*, 226.

65. "An Account of the Invasion of South Carolina by the French & Spaniards in the Month of August 1706," in *Records in the British Public Record Office Relating to South Carolina 1701–1710*, ed., A. S. Salley (Columbia: Historical Commission of the State of South Carolina, 1947), 161–65.

66. See Barlow Cumberland, *History of the Union Jack and Flags of the Empire* (W. Briggs, 1909), 281–82; and Chapin, *Privateer Ships*, 10–11.

67. "An Impartial Narrative of ye Late Invasion of So Carolina by ye French + Spaniards, in the Month August 1706" in *Records in the British Public Record Office*, 181–85.

68. "Massachusetts Documents, 1689–1692."

69. John Henry Edmonds, *Captain Thomas Pound* (Cambridge: John Wilson and Son, 1918), 31–32. Ironically, the same anti-Catholic anxieties and prejudices that led New Englanders to overthrow Andros also led them to limit their naval defenses to small provincial sloops. For the connection between antipopery and the Glorious Revolution and King William's War, see Stanwood, *Empire Reformed*.

70. The sloop *Resolution* was used by provincial forces early on, but in *Captain Thomas Pound*, Edmonds claims that "somehow or other [it] had got [sic] into private hands" (34). In *Privateer Ships* (96), Chapin claims that *Resolution* was sold by 1693.

71. Chapin, *Privateer Ships*, 96. Chapin contests Edmonds's argument that Pound's piracy was a mere cover for his political opposition to the Glorious Revolution.

72. Edmonds, *Captain Thomas Pound*, 33.

73. Alan Rogers, *Murder and the Death Penalty in Massachusetts* (Amherst: University of Massachusetts Press, 2008), 15–16.

74. Increase Mather, "A Vindication of New England," in *The Andros Tracts: Being a Collection of Pamphlets and Official Papers [. . .]*, vol. 6, ed. William Henry Whitmore (Boston: Prince Society, 1869), 37. See Edmonds, *Captain Thomas Pound*, for that historian's view that Pound was part of a larger farcical attempt by some colonial leaders to get the Massachusetts government to restore HMS *Rose*. No other historians have concurred with Edmonds's claims.

75. "The humble Address and Petition of the Governor and Councille and the Representatives of the Colony of the Massachusetts Bay; convened in General Court at Boston," in *Andros Tracts*, 43–44.

76. "Sir Edmund Andros' account of the force raised in the year 1688 for the defence of New England against the Indians."

77. Cooke and Oakes, "An Answer to Sr: Edmond Andros's Acco."

78. "At the Court at Whitehall, the 26th: Aprill 1690," in "New England patents and grants, 1690," Submission, Order, TNA, Kew, CO 5/905, 119–20.

79. In *Privateer Ships* (95–96), Chapin very simply remarks that Bradstreet's government refused to surrender *Mary* and had already sent *Speedwell* to England. When I examined the original documents however, I was struck with how they were rife with many documentary and logical inconsistencies. While the actors frequently omitting the names of the disputed vessels certainly added to the confusion, many of the sources simply don't supply a comprehensible narrative of the dispute. For instance, a year after King William ordered Bradstreet to give Governor Sloughter one of the sloops, Bradstreet claimed that Andros never had sloops built at the "publick charge" but had made use of one sloop (presumably *Mary*). Bradstreet alleged that the *Mary* had been jointly funded by New England governments before New York was added to the Dominion and had another sloop (presumably *Speedwell*) in construction when the revolution happened. Neither Andros, the New England agents, or Bradstreet made mention of the *Resolution* sloop—a third provincial sloop. Evidently, both *Speedwell* and *Resolution* were sent to England and sold by provincial agents. To add to this contradictory morass, Governor Sloughter wrote the king to inform him that he had received *Mary* but then immediately dispatched a vessel to seize *Mary* even after he admitted

acquiring the vessel. What is clear from the chaotic sources is that the only dominion-era vessel left in Boston as of 1691 was the sloop *Mary*, captained by John Alden.

For some of these sources, see "Letter concerning a sloop of war," May 8, 1691, Correspondence, TNA, Kew, CO 5/856, 173–74; "Account of moneys disbursed and paid for a new sloop built by John Cooke for their majesties service," May 8, 1691, Financial Document, TNA, Kew, CO 5/856, 164; "Exceptions to the Province Accot of John Phillips Esqr Late treasurer," in *The Acts and Resolves, Public and Private, of the Province of the Massachusetts Bay*, vol. 7 (Boston: Wright & Potter Printing, 1892), 409–10; and Increase Mather, "The Present State of the New English Affairs, 3 September 1689," in *Andros Tracts*, vol. III, 62.

80. *Calendar of Historical Manuscripts, in the Office of the Secretary of State*, part 2, ed. E. B. O'Callaghan (Albany: Weed, Parsons Printers, 1866), 226. Also, see "Archangel," at the Three Decks Age of Sail website/Naval History database.

81. Louise A. Breen, *Transgressing the Bounds: Subversive Enterprises among the Puritan Elite in Massachusetts, 1630–1692* (Cary: Oxford University Press, 2001), 203–8, ProQuest.

82. Paul Boyer and Stephen Nissenbaum, *Salem Possessed* (Cambridge: Harvard University Press, 1974), 32, JSTOR. For other treatments on Alden in the trials, see Benjamin C. Ray. *Satan and Salem : The Witch-Hunt Crisis of 1692* (Charlottesville: University of Virginia Press, 2015), 199–200; and John McWilliams, "Indian John and the Northern Tawnies," *New England Quarterly* 69, no. 4 (1996): 588–90.

83. Mary Beth Norton, *In the Devil's Snare: The Salem Witchcraft Crisis of 1692* (New York: Vintage Books, 2002), Kindle eBook edition, Loc. 3815.

84. Breen, *Transgressing the Bounds*, 205–8.

85. Norton, *In the Devil's Snare*, Loc. 3719–51.

86. Ray, *Satan and Salem*, 199–200; and Breen, *Transgressing the Bounds*, 208.

87. Council Minutes, Massachusetts, October 9, 1704, "Minutes of the Massachusetts council, Jun 1704–Mar 1705," Minutes, TNA, Kew, CO 5/789, 310–12, Adam Matthew, Colonial America Digital Database.

88. Council Minutes, Massachusetts, September 4, 1701, "Minutes of the council of the Massachusetts Bay from 30th May to 17th September 1701," Minutes, TNA, Kew, CO 5/788, 43, Adam Matthew, Colonial America Digital Database.

89. Minutes of Council [in Assembly] of Barbados, August 25–26, 1702, and Journal of Assembly of Barbados, August 25, 1702, in *CSP*, vol. 20, 1702, ed. Cecil Headlam, 548–66 (London: His Majesty's Stationery Office, 1912), British History Online.

90. Chapin, *Privateer Ships*, 249.

91. Minutes of Council in Assembly of Barbados, September 8, 1702; and Journal of the Assembly of Barbados, September 9, 1702, in *CSP*, vol. 20, 1702, 581–88, British History Online.

92. Minutes of Council in Assembly of Barbados, September 15, 1702, in *CSP*, vol. 20, 1702, 588–92, British History Online.

93. Chapin, *Privateer Ships*, 249.

94. Minutes of the Council in Assembly of Barbados, September 18, 1702, and Journal of Assembly of Barbados, September 19, 1702, in *CSP*, vol. 20, 1702, 592–99, British History Online.

95. Leach, *Roots of Conflict*, 235–43.
96. Denver Brunsman, *The Evil Necessity: British Naval Impressment in the Eighteenth-Century Atlantic World* (Charlottesville: University of Virginia Press, 2013), 107–12, ProQuest; and Satsuma, *Britain and Colonial Maritime War*, 134–46.
97. Miles, *Royal Navy*, 110–24, 137, 135–36; Tapley, *Province Galley*, 2; and "The Deposition of Capt Thomas Dobbins late Comander of their Majties Ship Nonsuch now Commander of their Majties Gally called the Province Gally [. . .]," in "Affadavits of various sailors and officials relating to the complaints of Jahleel Brenton and Richard Short against Sir William Phips," Legal Document, TNA, Kew, CO 5/858, 211, Adam Matthew, Colonial America Digital Database.
98. Nagel, *Empire and Interest*, 291.

CHAPTER 3

1. John Brewer, *The Sinews of Power: War, Money and the English State, 1688–1783* (London: Unwin Hyman, 1989), 139–40, JSTOR. Rodger, *Command of the Ocean*, 291–92.
2. Adrian Finucane, *The Temptations of Trade: Britain, Spain, and the Struggle for Empire* (Philadelphia: University of Pennsylvania Press, 2016), 71, ProQuest. For more on *guarda costas* and Spanish responses to Anglo-American smuggling and piracy, see G. Earl Sanders, "Counter-Contraband in Spanish America: Handicaps of the Governors in the Indies," *Americas* 34, no. 1 (1977): 59–72, JSTOR.
3. Richard Harding, *The Emergence of Britain's Global Naval Supremacy: The War of 1739–1748* (Woodbridge: Boydell and Brewer, 2010), 17–18, JSTOR.
4. Hanna, *Pirate Nests*, 366–67.
5. Peter Earle, *The Pirate Wars* (Macmillan, 2003), 158–61, Kindle eBook edition. There is considerable dispute about the proper periodization of the Golden Age of Piracy. For the sake of this study, the 1715–25 framework is most applicable. For more on the scholarly debate, see Hanna, "Well-Behaved Pirates Seldom Make History: A Reevaluation of the Golden age of English Piracy," in *The Sea in the Early Modern Era Essays in Honor of Robert C. Ritchie*, ed. Peter C. Mancall and Carole Shammas. (San Diego: University of California, 2015).
6. Bahar, *Storm of the Sea*, 160–62.
7. Ivers, *This Torrent of Indians*, 109.
8. Rodger, *Command of the Ocean*, 294–303. For the numbers and disposition of Royal Navy ships immediately following Queen Anne's War, see Clive Wilkinson, *The British Navy and the State in the Eighteenth Century* (London: Boydell & Brewer, 2004), 68–69.
9. McLaughlan, *Sloop of War*, 143–44.
10. Kinkel, *Disciplining the Empire*, 54–58. Kinkel's recent study contrasts starkly with Richard Harding's contention that by 1730 by "the apparent power of the Royal Navy in the Caribbean was seen to be so great that merchants wanted operations curbed to prevent the Spanish trade being destroyed." See Harding, *Seapower and Naval Warfare*, 189–92.
11. Eliga Gould, *Among the Powers of the Earth: The American Revolution and the Making of a New World Empire* (Cambridge: Harvard University Press, 2012), 35–36, ProQuest;

Wilson "Protecting Trade by Suppressing Pirates," 100–102. Also see Jeffers Lennox, *Homelands and Empires: Indigenous Spaces, Imperial Fictions, and Competition for Territory in Northeastern North America, 1690–1763* (Toronto: University of Toronto Press, 2017), 70–71, ProQuest.

12. A note must be made here about the fluidity of the names of various maritime actors in this period. As has been previously mentioned, British authorities often accused Spanish coast guards of outright piracy and some scholars have recently discussed how New England officials called Wabanaki mariners "pirates" to delegitimize their foes as mere criminals. Other historians have also categorized eighteenth-century piracy as one choice along a "continuum" of legal and illegal maritime activity. Such categorical fluidity was present among provincial naval forces as well, particularly in the West Indies. As during the previous two imperial wars, colonial officials referred to provincial naval operations with language varying from "private men of war" to "guard sloops" to "privateers" to "publick privateers." With the onset of Spanish *guarda costas* after Queen Anne's War, some colonial officials even called their own provincial guard vessels by derivations of that title. With strict definitions of privateering and piracy still very much up for debate in this period, it is unsurprising that Anglo-Americans continued to use many different names for their maritime defense options. As in the previous chapter, I argue there was a "provincial naval continuum" that ranged from state-funded and controlled warships to privateers with letters of marque. This is similar in many regards to Kevin P. McDonald's analysis of piracy as one expression of a "spectrum" of criminal activity in the eighteenth-century Caribbean. See "Sailors from the Woods: Logwood Cutting and the Spectrum of Piracy," in Head, *Golden Age of Piracy*, 52–54.

13. David Wilson makes the point that South Carolina (like Rhode Island and Pennsylvania) was not guaranteed a Royal Navy ship as a proprietary colony. Massachusetts on the other hand had a station ship due to its status as a royal colony. While this did not guarantee political stability (as we have seen in the last chapter), this would make a major difference when it came to antipirate expeditions and their political ramifications. See Wilson, "Protecting Trade," 98–99.

14. *Boston News-Letter*, May 27, 1717, 2, Readex, America's Historical Newspapers.

15. Legislature Minutes, Massachusetts, June 4, 1717, in *Journals of the House of Representatives of Massachusetts, 1715–1717* (Boston: Massachusetts Historical Society, 1919), 186–87.

16. John Grenier, *The Far Reaches of Empire: War in Nova Scotia, 1710–1760* (Norman: University of Oklahoma, 2008), 12–15.

17. Bahar, *Storm of the Sea*, 160–62.

18. Lipman, *Saltwater Frontier*, 80–82.

19. Bahar, *Storm of the Sea*, 171–72.

20. Bahar, *Storm of the Sea*, 127.

21. "Boston, April 15." *American Weekly Mercury* (Philadelphia), May 2, 1723, 2, Readex, America's Historical Newspapers.

22. Grenier, *Far Reaches*, 40–45.

23. *Boston News-Letter*, September 19, 1720, 4, Readex, America's Historical Newspapers. And Governor Richard Philipps to the Council of Trade and Plantations,

September 27, 1720 (and attachments), in *CSP*, vol. 32, *1720–1721*, ed. Cecil Headlam, 144–65 (London: His Majesty's Stationery Office, 1933), British History Online.

24. Governor Richard Philipps to the Council of Trade and Plantations, August [6?], 1720, in *CSP*, vol. 32, *1720–1721*, ed. Cecil Headlam, 77–97 (London: His Majesty's Stationery Office, 1933), British History Online.

25. Governor Richard Philipps to the Council of Trade and Plantations, September 27, 1720, 144–65.

26. Council of Trade and Plantations to Mr. Secretary Craggs, December 14, 1720 (No. 322), in *CSP*, vol. 32, *1720–1721*, ed. Cecil Headlam, 212–28 (London: His Majesty's Stationery Office, 1933), British History Online.

27. Governor Richard Philipps to the Council of Trade and Plantations, August 16, 1721, in *CSP*, vol. 32, *1720–1721*, ed. Cecil Headlam, 388–402 (London: His Majesty's Stationery Office, 1933), British History Online.

28. "Petition of Governor Philipps to the King" in *CSP*, vol. 34, *1724–1725*, ed. Cecil Headlam and Arthur Percival Newton, 86–105 (London: His Majesty's Stationery Office, 1936), British History Online. Also see Hitchings, "Guarding the New England Coast," 58–61; and Lennox, *Homelands*, 70–71. Some sources (such as Hitchings above) call the *William Augustus* a "schooner," and contemporary sources call it a "sloop." During the interwar period, the Royal Navy operated some schooner-rigged vessels that were called "sloops," and it is likely the *William Augustus* fell into this category. See McLaughlan, *Sloop of War*, 143.

29. J. Burchett to the Lords of the Treasury, December 19, 1723 (No. 49), in "Volume 248: July 3–December 31, 1724," in *Calendar of Treasury Papers*, vol. 6, *1720–1728*, ed. Joseph Reddington, 276–98 (London: Her Majesty's Stationery Office, 1889), British History Online.

30. In *Nova Scotia and the Royal Navy, 1713–1766*, unpublished diss., Queen's University, 1973, microfilm, 25–30, W. A. B. Douglas notes the importance of Royal Navy intervention in the area after 1710. Nevertheless, more recent scholars such as Lennox (*Homelands*) and Grenier (*Far Reaches*) have underlined the importance of provincial military forces in securing the region during this tense era.

31. Hanna, *Pirate Nests*, 377. A special thanks to Dr. Eliga Gould for bringing my attention to this example.

32. Bahar, *Storm of the Sea*, 183. For a primary source detailing some of the ways in which Anglo-Americans met their match when fighting Wabanaki mariners, see Samuel Penhallow, *The History of the Wars of New-England with the Eastern Indians* (1726; repr., Boston: Oscar H. Harpel, Chestnut Street, 1859), 101–2.

33. Council Minutes, Massachusetts, June 6, 1722, in "Minutes of Council of the Province of the Massachusetts Bay 2 Mar–20 Aug 1722," Minutes, TNA, Kew, CO 5/794, 51–52, Adam Matthew Colonial America Digital Database.

34. *Boston News-Letter*, June 11, 1722, 2, Readex, America's Historical Newspapers.

35. "Order on Committees Report About Ye Defence of the Coast Agst Pyrates" and "Vote for Encouraging the Prosecution of Ye Pyrates," in *The Acts and Resolves, Public and Private, of the Province of the Massachusetts Bay*, vol. 10, *Resolves, Etc. 1720–1726* (Boston: Wright & Potter Printing, State Printers, 1902), 163–64.

36. Paper money was a controversial subject in this era, and its larger implications reach beyond the scope of this chapter. For a more thorough discussion of both South Carolina's and New England's concurrent battles over paper money, see M. Eugene Sirmans, *Colonial South Carolina: A Political History 1663–1763* (Chapel Hill: University of North Carolina Press, 1966); and Margaret Ellen Newell, *From Dependency to Independence: Economic Revolution in Colonial New England* (Cornell University Press, 2015). Aside from paying sailors who captured pirates (who were also eligible for royal rewards), the colony reimbursed vessels impressed by the Royal Navy to hunt pirates with paper money. See "Chapter 313. Resolve Allowing £31 to Nathl. Masters," in *Acts and Resolves*, vol. 10, 388.

37. D. B. Quinn, "CUMINGS, ARCHIBALD," in *Dictionary of Canadian Biography*, vol. 2 (University of Toronto/Université Laval, 2003); and Archibald Cumings to William Popple, June 20, 1722, in "Letters from Samuel Shute and Archibald Cumings to the Board of Trade, enclosing Council Minutes," Correspondence, Minutes, TNA, Kew, CO 5/868, Adam Matthew, Colonial America Digital Database.

38. Legislature Minutes, Massachusetts, June 28, 1722, in *Journals of the House of Representatives of Massachusetts*, vol. 4, 1722–1723 (Boston Massachusetts Historical Society, 1923), 54–58; and Council Minutes, Massachusetts, June 29, 1722, in "Minutes of Council of the Province of the Massachusetts Bay 2 Mar–20 Aug 1722," Minutes, TNA, Kew, CO 5/794, Adam Matthew, Colonial America Digital Database.

39. Council Minutes, Massachusetts, July 25, 1722, in "Minutes of Council of the Province of the Massachusetts Bay 2 Mar–20 Aug 1722." Minutes, TNA, Kew, CO 5/794, 59–60, Adam Matthew Colonial America Digital Database.

40. "Vote for Fitting out Two Shallops Against the Indians," August 18, 1722, in *The Acts and Resolves, Public and Private, of the Province of the Massachusetts Bay*, vol. 10, *Resolves, Etc., 1720–1726* (Boston: Wright & Potter Printing, 1902), 217.

41. Legislature Minutes, Massachusetts, June 27–28, 1722, in *Journals of the House of Representatives of Massachusetts*, vol. 4, 1722–1723 (Boston Massachusetts Historical Society, 1923), 51, 55.

42. Emma Lewis Coleman, *New England Captives Carried to Canada Between 1677 and 1760 during the French and Indian Wars*, vol. 1 (Heritage Books, 2008), 99.

43. Bahar, "People of the Dawn, People of the Door: Indian Pirates and the Violent Theft of an Atlantic World," *Journal of American History* 101, no. 2 (September 2014): 417–18; Bahar, *Storm of the Sea*, 172–78; Council Minutes, Massachusetts, June 25–26, 1722, in "Minutes of Council of the Province of the Massachusetts Bay 2 Mar–20 Aug 1722," Minutes, TNA, Kew, CO 5/794, 53–54, Adam Matthew, Colonial America Digital Database. For the effects of Wabanaki victories on provincial morale, see John Minot to William Dummer, July 16, 1724, in *Letters of Colonel Thomas Westbrook and Others Relative to Indian Affairs in Maine 1722–1726*, ed. William Blake Trask (Boston: G. E. Littlefield, 1901), 64–65.

44. Grenier, *Far Reaches*, 57–69. Grenier's view of the peace terms is more positive than Bahar's. Grenier suggests that New Englanders succeeded in achieving favorable terms, while Bahar has recently made the case that Wabanaki leaders overwhelmed Anglo-American shipping and made peace from a stronger position. See Bahar, *Storm of the Sea*, 183–85.

45. See various transport vessel muster rolls in Trask, *Letters of Colonel Thomas Westbrook*, 171–79; and *American Weekly Mercury* (Philadelphia), April 30, 1724, 2, Readex, America's Historical Newspapers.

46. Council Minutes, Massachusetts, June 28, 1726. in "Minutes of the Council of Massachusetts Bay," Minutes, TNA, Kew, CO 5/797, 126, Adam Matthew, Colonial America Digital Database.

47. "Memorial of the Lt. Governor and Council of the Massachusetts Bay to the King," July 8, 1726, in *CSP*, vol. 35, *1726–1727*, ed. Cecil Headlam and Arthur Percival Newton, 96–115 (London: His Majesty's Stationery Office, 1936), British History Online.

48. Council Minutes, Massachusetts, June 29, 1726, CO 5/797, 126.

49. "Memorial of the Lt. Governor."

50. Logbook of HMS *Sheerness*, ADM 51/898, TNA.

51. Douglas Edward Leach, *Roots of Conflict: British Armed Forces and Colonial Americans, 1677–1763* (Chapel Hill: The University of North Carolina Press, 1986), p. 141, Kindle eBook Version.

52. Legislature Minutes, Massachusetts, August 27, 1726, in *Journals of the House of Representatives of Massachusetts, 1726–1727* (Boston: Massachusetts Historical Society, 1926), 103.

53. In *Empire and Interest*, Kurt Nagel makes the case that while New York and Massachusetts faced Indian conflicts after Queen Anne's War, "neither colony was as harried as the southern frontier" (295). While both Massachusetts and South Carolina dealt with piratical and Native maritime attacks throughout the late 1710s to early 1720s, the costs of provincial naval defense were more dramatic for South Carolina thanks to its near destruction during the Yamasee War.

54. Steven J. Oatis, *A Colonial Complex: South Carolina's Frontiers in the Era of the Yamasee War, 1680–1730* (Lincoln: University of Nebraska Press, 2004), 114–26, EBSCOhost. It should be noted that the Yamasee pitted South Carolinians against other tribes as well, including the Apachiolas (141). For other causes of the conflict and for more information on the region-wide causes of the war, see Allan Gallay, *The Indian Slave Trade: The Rise of the English Empire in the American South, 1670–1717* (New Haven: Yale University Press, 2002), 329–37, ProQuest; and Sirmans, *Colonial South Carolina*, 114–15.

55. Lynn B. Harris, *Patroons and Periaguas: Enslaved Watermen and Watercraft of the Lowcountry* (Columbia: University of South Carolina Press, 2014), 2–25, EBSCOhost. Alexander Y. Sweeney, "Cultural Continuity and Change: Archaeological Research at Yamasee Primary Towns in South Carolina," in *The Yamasee Indians: From Florida to South Carolina*, ed. Denise Bossy (Lincoln: University of Nebraska Press, 2018), 121, JSTOR.

56. "An Act for the Better Security of That Parte of the Province of Carolina ... Against Any Hostile Invasions and Attempts by Sea or Land, Which the Neighbouring Spaniard or Other Enemy May Make Upon the Same," in *The Statutes at Large of South Carolina*, vol. 2, from 1682–1716, ed. Thomas Cooper (Columbia, SC: A. S. Johnston, 1837), 11–13, and "An Act for Appointing Two Scout Canoes," in *South Carolina Statutes at Large*, vol. 2, ed. Thomas Cooper (Columbia, SC: A. S. Johnston, 1837), 607–8.

57. Oatis, *Colonial Complex*, 144–45; and Ivers, *This Torrent*, 64–67.

58. Ivers, *This Torrent*, 107–9.

59. Francis Le Jau to the Society of the Propagation of the Gospel, qtd. in Edgar Legare Pennington, "The South Carolina Indian War of 1715, as Seen by the Clergymen," *South Carolina Historical and Genealogical Magazine* 32, no. 4 (October 1931): 257, JSTOR.

60. Ivers, *This Torrent*, 82–83.

61. Council Minutes, Massachusetts, June 7, 1715, in "Massachusetts: Minutes of Council 21 Mar–11 Oct 1715," Minutes, TNA, Kew, CO 5/792, 111, Adam Matthew, Colonial America Digital Database; and Governor Joseph Dudley to the Council of Trade and Plantations, July 31, 1715, in *CSP*, vol. 28, *1714–15*, ed. Cecil Headlam, 235–53 (London: His Majesty's Stationery Office, 1928), British History Online.

62. Captain Samuel Mead to Admiralty Secretary Josiah Burchett, December 27, 1715, ADM 1/2095, TNA.

63. Oatis, *Colonial Complex*, 162–63.

64. Lords Proprietors of Carolina to the Council of Trade and Plantations (No. 517), in *CSP*, vol. 28, ed. Cecil Headlam, 215–35 (London: His Majesty's Stationery Office, 1928), British History Online.

65. Oatis, *Colonial Complex*, 180–82.

66. Sirmans, *Colonial South Carolina*, 114–15. For more on post-1717 tensions with the Yamasee, see Ivers, *This Torrent*, chapters 14–17.

67. Governor Robert Johnson to the Council of Trade and Plantations, January 12, 1720, in *CSP*, vol. 31, *1719–1720*, ed. Cecil Headlam, 293–311 (London: His Majesty's Stationery Office, 1933), British History Online.

68. Carl E. Swanson, "'The Unspeakable Calamity This Poor Province Suffers from Pyrats': South Carolina and the Golden Age of Piracy," *Northern Mariner/Le Marin Du Nord* 21, no. 2 (2011): 119–23.

69. Governor and Council of South Carolina to the Council of Trade and Plantations, October 21, 1718, in *CSP*, vol. 30, *1717–1718*, ed. Cecil Headlam, 359–81 (London, His Majesty's Stationery Office 1930), British History Online.

70. Robert Johnson to Colonel William Rhett, September 4, 1718, in "South Carolina Probate Records, Bound Volumes, 1671–1977," Charleston Wills, 1716–1721, South Carolina Department of Archives and History, Columbia, FamilySearch website.

71. Swanson, "Unspeakable Calamity," 129–30.

72. Charles Johnson, *A General History of the Pirates, from Their First Rise and Settlement in the Island of Providence, to the Present Time* (London: T. Warner, 1724), 98–100; and Nicholas Trott, "The Lord Chief Justice's Speech, upon his pronouncing Sentence on Major Stede Bonnet," in Johnson, *General History*, 107–8.

73. *London Gazette*, September 14–17, 1717. While royal rewards for distant pirate captures sounded generous to enterprising colonial mariners, they inevitably ran into years of red tape. Records from Calendar of Treasury Papers from 1723 indicated that the provincial sailors who hunted down Stede Bonnet were still seeking payment five years after his capture! See the entry for August 27 in "Volume 244: July 1–December 30, 1723," in *Calendar of Treasury Papers*, vol. 6, *1720–1728*, ed. Joseph Redington, 218–37 (London, 1889), British History Online.

74. These are highlights from a larger court transcript that can be found in *Records*

of the South Carolina Court of Admiralty, 1716–1732, National Archives, Washington, DC, n.d., microfilm, 306–90. Accessed at Charleston County Library, South Carolina, History Room. I would like to express my sincere thanks to Dr. Nic Butler and the staff there for their assistance.

75. Richard P. Sherman, *Robert Johnson: Proprietary & Royal Governor* (Columbia: University of South Carolina Press, 1966), 34–35.

76. Johnson, *General History*, 2: 325–28.

77. Jabbs, *South Carolina Colonial Militia*, 280–81, 340–45. Historians have recently challenged the notion that the primary issue with the Lords Proprietors was ineffective military protection. For instance, Hanno T. Scheerer has made the case that "constitutional issues rather than questions of defense lay at the core of the quarrels between the Carolina proprietors and their settlers." See "The Proprietors Can't Undertake for What They Will Do: A Political Interpretation of the South Carolina Revolution of 1719," in *Creating and Contesting Carolina: Proprietary Histories*, ed. Michelle LeMaster and Bradford J. Woods (Columbia: University of South Carolina Press, 2013).

78. Harding, *Seapower and Naval Warfare*, 189–90.

79. Jabbs, *South Carolina Colonial Militia*, 349–54; and "The Humble Address of the Representatives of the Inhabitants of the Said Province, Now Conven'd at Charles Town," in Francis Yonge, *A Narrative of the Proceedings of the People of South-Carolina, in the Year 1719: And of the True Causes and Motives that Induced Them to Renounce Their Obedience to the Lords Proprietors, as Their Governors, and to Put Themselves Under the Immediate Government of the Crown*, vol. 1 (London: 1726), 31–32.

80. Yonge, *Narrative*, 8–10.

81. Charles Christopher Crittenden, "The Surrender of the Charter of Carolina," *North Carolina Historical Review* 1, no. 4 (October 1924): 393–95, JSTOR; and Log Book of HMS *Flamborough*, TNA, ADM 51/357.

82. Yonge, *Narrative*, 40.

83. Verner W. Crane, *The Southern Frontier, 1670–1732* (Durham: Duke University Press, 1928), 187–91.

84. "An Answer to the Queries sent by the Honble the Lords Comissioners of trade and plantations relating to the State of South Carolina," in William R. Coe Collection, 1699–1741, South Carolina Historical Society (hereafter referred to as SCHS); and Council of Trade and Plantations to the Lords Justice, September 23, 1720, in *CSP*, vol. 32, 1720–1721, ed. Cecil Headlam, 144–65 (London: His Majesty's Stationery Office, 1933), British History Online. It is important to note that the Board of Trade did not initiate the plans for the southern fort on the Altamaha, but this plan was developed by South Carolina agent/war veteran John Barnwell. See Sirmans, *Colonial South Carolina*, 134–36.

85. Ivers, *This Torrent*, 175–76. Also see "An Agreement made by order of His Excellency Francis Nicholson . . . By Coll: John Barnwell . . . With Jonathan Collings Commander of the Sloop Jonathan and Sarah [. . .]," June 9, 1721, in "Letters and papers relating to the landing of his Majesty's Independent Company now in south Carolina, and likewise concerning Colonel Barnwell's going to Altamaha river in order to build a fort there," Correspondence, Submission, Warrant, TNA, Kew, CO 5/358, Adam Matthew, Colonial America Digital Database.

86. Sirmans, *Colonial South Carolina*, 30–33.

87. Harding, *Emergence*, 17–18; Ivers, *This Torrent*, 187–88; "Extract of a Letter from a Merchant in Carolina, to his Friend in London, Dated, Sept. 14, 1727," *Boston Gazette*, February 5, 1728, 1, Readex, America's Historical Newspapers. For more information on proprietary attempts to retake the colony, see Sirmans, *Colonial South Carolina*, 153–54.

88. Dr. Nic Butler, "Anson and the Privateer Emergency of 1727," unpublished MSS. Dr. Butler, a leading archivist at the Charleston Archives and author of an upcoming monograph on Admiral George Anson, was kind enough to share some of his discoveries from the admiralty archives at Kew with me, and his insight has been invaluable for this chapter. For more on impressment policies at the time, see Brunsman, *Evil Necessity*, 109–11, ProQuest eBook. There were other cases where Royal Navy officials depended on provincial vessels from South Carolina to assist them in various tasks in the 1720s. For example of cooperation between Royal Navy forces and provincial navy scout boats in surveying the coast, see James Gascoigne, qtd. in W. E. May, "The Surveying Commission of Alborough, 1728–1734," *American Neptune* 21, no. 4 (October 1961): 264.

89. "An act for carrying on several expeditions against the Indian and other enemies and for defraying the charge thereof," September 30, 1727, Legislation, TNA, Kew, CO 5/387, 292, Adam Matthew, Colonial America Digital Database. While this law called *Palmer* a "sloop of war," President Middleton's commission to Captain Mountjoy called his vessel a "Private Sloop of War" (a term often applied to privateers), limited the territories in which he could patrol, and cautioned Mountjoy to "pay due regard" to any Royal Navy vessel that might stop him. In an April 1727 petition asking for monetary compensation for various expenses, Mountjoy himself called his vessel a "Guard de Coste," a term fairly similar to the Spanish *guarda costas* then so prevalent throughout the West Indies and continental coastline. It seems that Mountjoy's mission straddled the line between traditional "letter of marque" privateering and a provincial naval expedition. See "Copy of Captain Montjoys commission and instructions," Commission, Order, TNA, Kew, CO 5/387, 317; and Journal of the Commons House of Assembly, April 4, 1728, Minutes, TNA, Kew, CO 5/430, 74–75, Adam Matthew, Colonial America Digital Database.

90. Sirmans, *Colonial South Carolina*, 151–57; and Joseph Albert Ernst, *Money and Politics in America 1755–1775* (Chapel Hill, University of North Carolina, 1973), 30–31.

91. *American Weekly Mercury* (Philadelphia), January 9, 1728, 2, Readex, America's Historical Newspapers.

92. Ivers, *This Torrent*, 189–97; and Oatis, *Colonial Complex*, 283; "Journal of the Commons House of Assembly," September 1, 1727, Minutes, TNA, Kew, CO 5/429, 225–27, Adam Matthew, Colonial America Digital Database.

93. Oatis, *Colonial Complex*, 284–85.

94. Daniel A. Baugh, ed. *Naval Administration, 1715–1750* (London: Navy Records Society, 1977), 328–30.

95. Earle, *Pirate Wars*, 183, Kindle eBook edition.

96. Anglo-American officials in North America also deployed privateers during the

two short conflicts with Spain in this period, but West Indian governments seemed to deploy "private men of war" far more frequently when fighting *guarda costas* and pirates than their continental brethren. In *Command of the Ocean* (228–29), N. A. M. Rodger discusses New England privateers active in the West Indies during the War of the Quadruple Alliance.

97. Wilson, "Protecting Trade," 104.

98. Satsuma, *Britain and Colonial Maritime War*, 230–43; and Kinkel, *Disciplining the Empire*, 62–66.

99. Governor Henry Worsley to the Council of Trade and Plantations, March 26, 1726, in *CSP*, vol. 33, 1722–1723, ed. Cecil Headlam, 221–38 (London: His Majesty's Stationery Office, 1934), British History Online.

100. [Charles Delafaye?] to Governor Henry Worsley, August 1723, *CSP*, vol. 33, 1722–1723, ed. Cecil Headlam, 221–38 (London: His Majesty's Stationery Office, 1934), British History Online.

101. E. T. Fox, "Jacobitism and the 'Golden Age' of Piracy, 1715–1725," *International Journal of Maritime History* 22, no. 2 (December 2010): 282–84. And Colin Woodward, *The Republic of Pirates: Being the True and Surprising Story of the Caribbean Pirates and the Man Who Brought Them Down* (Orlando: Harcourt, 2007), 100–103, Kindle eBook.

102. It is important to note that Hamilton did actually have experience with the deployment of provincial naval forces during Queen Anne's War. See Council and Assembly Minutes, Jamaica, September 29, 1712, TNA, CO 140/11.

103. Council and Assembly Minutes, Jamaica, September 7, 1716, TNA, CO 140/13, 819–27.

104. Lord Archibald Hamilton, *An Answer to an Anonymous Libel, Entitled, Articles Exhibited Against Lord Archibald Hamilton, Late Governour of Jamaica: With Sundry Depositions and Proofs Relating to the Same* (London, 1718), 46.

105. Hamilton, *Answer*, 73; and Fox, "Jacobitism," 283–84.

106. Wilson, "Protecting Trade," 96–97. See Earle, *Pirate Wars*, 181–90, for another scholar's view of the problems with Royal Navy pirate patrols, the 1717 Pardon, etc.

107. Woodward, *Republic*, 264–71.

108. Woodes Rogers to the Council of Trade and Plantations, October 31, 1718, *CSP*, vol. 30, 1717–1718. ed. Cecil Headlam, 359–81 (London: His Majesty's Stationery Office, 1930), British History Online.

109. Hanna, *Pirate Nests*, 370.

110. Earle, *Pirate Wars*, 196; Wilson "Protecting Trade," 97–99.

111. Governor Sir Nathaniel Lawes to the Council of Trade and Plantations, January 31, 1719, *CSP*, vol. 31, 1719–1720, ed. Cecil Headlam, 1–21 (London: His Majesty's Stationery Office, 1933), British History Online.

112. Council and Council in Assembly Minutes, Jamaica, January 8, 1718/19, TNA, CO 140/16.

113. Lawes to Council of Trade and Plantations, January 31, 1719.

114. Lawes to Council of Trade and Plantations, March 24, 1720, in *CSP*, vol. 31, 1719–1720, ed. Cecil Headlam, 45–66 (London: His Majesty's Stationery Office, 1933), British History Online.

115. Wilson, "Protecting Trade," 98.

116. Lawes to Council of Trade and Plantations, November 13, 1720, in *CSP*, vol. 32, *1720–1*, ed. Cecil Headlam, 187–95 (London: His Majesty's Stationery Office, 1933), British History Online.

117. Steven C. Hahn, "The Atlantic Odyssey of Richard Tookerman: Gentleman of South Carolina, Pirate of Jamaica, and Litigant before the King's Bench," *Early American Studies* 15, no. 3 (Summer 2017): 539–90, JSTOR.

118. "Extract of Letter from Capt. Vernon to Mr. Burchett, 7 Nov. 1720" and "Copy of Act of Jamaica for Fitting out Sloops for Guarding the Coasts etc.," in *CSP*, vol. 32, *1720–1721*, ed. Cecil Headlam, 329–46 (London: His Majesty's Stationery Office, 1933), British History Online.

119. Governor Sir Nathaniel Lawes to the Council of Trade and Plantations, April 20, 1721, in *CSP*, vol. 31, *1720–1721*, ed. Cecil Headlam, 281–97 (London: His Majesty's Stationery Office, 1933), British History Online.

120. Letter from the Duke of Portland [...], February 8, 1724/5, TNA CO 137/16, 65–77.

121. Governor the Duke of Portland to the Council of Trade and Plantations, February 8, 1724/5, in *CSP*, vol. 34, *1724–1725*, ed. Cecil Headlam and Arthur Percival Newton, 320–35 (London: His Majesty's Stationery Office, 1936), British History Online.

122. *The State of the Island of Jamaica. Chiefly in Relation to its Commerce and the Conduct of the Spaniards in the West-Indies* (London: H. Whitridge, 1726), 24–27; and Governor the Duke of Portland to the Council of Trade and Plantations, June 1, 1726, in *CSP*, vol. 35, *1726–1727*, Cecil Headlam and Arthur Percival Newton, 76–94 (London: His Majesty's Stationery Office, 1936), British History Online.

123. Harding, *Emergence*, 17–18.

124. Kinkel, *Disciplining the Empire*, 66–67.

125. Council Minutes, Jamaica, April 7–9, 1729, TNA, CO 140/21.

126. Harding, *Emergence*, 17–19.

127. Bruce P. Lenman, *Britain's Colonial Wars, 1688–1783* (New York: Routledge, 2001), no page numbers given, Google Play eBook.

128. *New-England Weekly Journal* (Boston), April 30, 1733, 2, Readex, America's Historical Newspapers.

129. For more information on this case, the involvement of the Royal Navy, and the Spanish reaction, see Robert Hunter's letter to the Council of Trade and Plantations and its inserts in America and West Indies: March 1733, 16–31, in *CSP*, vol. 40, *1733*, ed. Cecil Headlam and Arthur Percival Newton, 51–69 (London: His Majesty's Stationery Office, 1939), British History Online.

130. "Extract, of a Letter from St. Christopher's, March 17. 1734, 5," *Boston News-Letter*, July 3, 1735, 2, Readex, America's Historical Newspapers.

CHAPTER 4

1. For the sake of simplicity, I have chosen to use the War of Jenkins' Ear as a catch-all term for the wars with the French and Spanish that occurred between 1739 and 1748, though the war with the French is technically known as King George's War.

Names for the European theaters of these wars are the War of the Austrian Succession and the War of the League of Augsburg.

2. Richard Pares, *War and Trade in the West Indies, 1739–1763* (London: Frank Cass, 1963), 29–64; Kinkel, *Disciplining the Empire*, 84–85; Satsuma, *Britain and Colonial Maritime War*, 220–21.

3. Ivers, *British Drums*, 7–11, 53–61. Ivers was one of the first scholars since Chapin to use the term "Provincial navy" to describe a colonial fleet.

4. Trevor R. Reese, "Georgia in Anglo-Spanish Diplomacy, 1736–1739," *William and Mary Quarterly* 15, no. 2 (April 1958): 170–78, JSTOR.

5. Daniel A. Baugh, *British Naval Administration in the Age of Walpole* (Princeton: Princeton University Press, 1965), 27.

6. Baugh, *Naval Administration*, 247; Harding, *Emergence*, 39–40, 52–53.

7. Christian Buchet, Anita Higgie, and Michael Duffy, *The British Navy, Economy and Society in the Seven Years War* (Woodbridge, Suffolk; Rochester, NY: Boydell & Brewer, 2013), 142–51, JSTOR.

8. The extent of this professionalization is controversial. Most historians, ranging from Daniel Baugh to Richard Harding, point to naval professionalization as a turning point for the Royal Navy's command structure, but Sarah Kinkel has recently expanded upon this view and attributed this professionalization to the rise of "Authoritarian" Whiggism in the British Admiralty. See Baugh, *Naval Administration*, 502–3; Harding, *Emergence*, 305–7; Kinkel, *Disciplining the Empire*, 90–91, 101–4, 117–20.

9. Swanson, *Predators and Prizes*, 29–30, 50–52.

10. For the wide variety of historians who have also noticed this increase during this era, see Howard Chapin, *Privateering in King George's War, 1739–1748* (Providence: E. A. Johnson, 1928), particularly chapter 3; and Chapin, *The Tartar: The Armed Sloop of the Colony of Rhode Island in King George's War* (Providence: Society of Colonial Wars, 1922) (see 11–12 for the *Tartar*'s journey to Cuba). See further discussion of Chapin's differentiation between provincial navies and privateers in chapter 2 of this book. H. W. Richmond: *The Navy in the War of 1739–48*, vol. 3 (Cambridge: Cambridge University Press, 1920), 273–74; and W. A. B. Douglas, "Sea Militia of Nova Scotia," 23–25. Swanson, *Predators and Prizes*, 50–52.

11. Ian K. Steele, *The English Atlantic, 1675–1740: An Exploration of Communication and Community* (New York: Oxford University Press, 1986), 93, EPUB; and *South Carolina Gazette* (Charles Town), May 30, 1743, Accessible Archives.

12. For examples of scholarly studies of this political gulf, see Greene, *Peripheries and Center*; Nagel, *Empire and Interest*, David Armitage. *The Ideological Origins of the British Empire* (Cambridge: Cambridge University Press, 2000), 190–200, ProQuest; and Gould, *Persistence of Empire*, 50–70.

13. W. A. B. Douglas's characterization of the "special relationship" between provincial vessels and the Royal Navy in securing the Nova Scotia coastline between 1745 and 1755 has been useful here. Throughout his unpublished dissertation on Nova Scotia's relationship with the Royal Navy, Douglas highlights the importance of both New England and Nova Scotian provincial vessels, especially in light of various strategic oversights by the Admiralty and Royal Navy. With this chapter, I expand upon Douglas's

findings and extend the notion of this "special relationship" to the rest of British Atlantic world. See Douglas, *Nova Scotia and the Royal Navy, 1713–1766*, 473.

14. Robert Dinwiddie, "Report to the Board of Trade, April 1740," qtd. in Kenneth Morgan, "Robert Dinwiddie's Reports on the British American Colonies," *William and Mary Quarterly*, 3rd ser., 65, no. 2 (April 2008): 305–46, JSTOR.

15. One excellent example of this growing awareness can be found in an April 1740 pamphlet by Irish newsman and printer George Faulkner. Faulkner argued that the American colonies were "so well peopled, and have such a Number of Ships and Sailors, that they are both able and willing to put out 40 or 50 large Ships of Force at their own Expence." He maintained that if governments from New England to the West Indies should provide over forty galleys "built in the Nature of the French or Spanish Gallies, their Men exercised to Arms as our Foot are," they could capture St. Augustine and other Spanish ports. Faulkner contended that these swift oared vessels—with roots in Greco-Roman navies of antiquity—would be suitable for action in the West Indies as they had been in the Mediterranean for millennia and that provincial naval efforts would give the Royal Navy more room to operate elsewhere. Faulkner would later claim that "What gives us the greater Certainty of Success in this War, is, the great Strength and vast Trade our Plantations in America have acquired since the last War" but did admit royal military assistance would still be necessary to some extent. See George Faulkner, *The Present State of the Revenues and Forces by Sea and Land Of France and Spain, Compared with those of Great Britain [. . .]* (Dublin: George Faulkner, 1740), 23–33. Faulkner was almost certainly inspired by France's Galley Corps—a force independent of the French Royal Navy. French galleys were sleek, swift, lightly armed, and depended on mixed crews of both volunteers and convicts. See Rif Winfield and Stephen S. Roberts, *French Warships in the Age of Sail, 1626–1786: Design, Construction, Careers and Fates* (South Yorkshire: Seaforth Publishing, 2017), 372–74.

16. Peter Warren to George Anson, April 2, 1745, in *The Royal Navy and North America: The Warren Papers, 1736–1752*, ed Julian Gwyn (Navy Records Society, 1973), 74. In his article "Sea Militia," historian W. A. B. Douglas emphasizes Warren's unique support for provincial forces helping to set a "precedent" for future "sea militias" in Canada (25–26).

17. John Tate Lanning, "The American Colonies in the Preliminaries of the War of Jenkins' Ear," *The Georgia Historical Quarterly* Vol. 11, No. 2 (June 1927), 140–41, JSTOR.

18. Council minutes, Pennsylvania, June 14, 1748, in *Minutes of the Provincial Council of Pennsylvania, From the Organization to the Termination of the Proprietary Government*, vol. 5 (Harrisburg, Theo. Fenn, 1851), 277–80, Google Play Ebook. Also see Swanson, *Predators and Prizes*, 148.

19. For an excellent description of the men sailing the scout boats of Oglethorpe's navy, see Francis Moore, *A Voyage to Georgia, Begun in the Year 1735. Containing, an Account of the Settling the Town of Frederica, . . . With the Rules and Orders . . . for that Settlement [. . .]* (London: Jacob Robinson, 1744).

20. Ivers, *British Drums*, 58–62.

21. Richard S. Dunn, "The Trustees of Georgia and the House of Commons, 1732–1752," *William and Mary Quarterly* 11, no. 4 (October 1954): 551–59, JSTOR.

22. The British government would have been very well aware of the lengths Oglethorpe went to to provide local naval forces. The widely-read *Gentleman's Magazine* covered his exploits in great detail. See *Gentleman's Magazine: And Historical Chronicle*, vol. 10, *For the Year M.DCCXL* (London: Edw. Cave, 1740), 139. Also see Ivers, *British Drums*, 62–63 and Trevor R. Reese, "Colonial Georgia in British Policy, 1732–1756," unpublished diss., University of London, 1955, 182–84.

23. "Instructions for Governor Oglethorpe," October 9, 1739, Order, TNA, Kew, CO 5/654, 225–31, Adam Matthew, Colonial America Digital Database; and Rodney E. Baine, "General James Oglethorpe and the Expedition against St. Augustine," *Georgia Historical Quarterly* 84, no. 2 (Summer 2000): 201, JSTOR. This latter source has an excellent overview of the historiography of the expedition.

24. Baine, "General James Oglethorpe," 208–9.

25. *Pennsylvania Gazette* (Philadelphia), May 1, 1740, 2, Readex, America's Historical Newspapers.

26. "Establishment of a company of highland foot etc," Military Document, TNA, Kew, CO 5/654, 273–75, Adam Matthew, Colonial America Digital Database.

27. *Report of the Committee Appointed by the General Assembly of South Carolina in 1740 on the St. Augustine Expedition Under General Oglethorpe, Collections of the Historical Society, of South Carolina*, vol. 4 (Charleston: Walker, Evans & Cogswell, 1887), 126.

28. Major James P. Herson, *A Joint Opportunity Gone Awry: The 1740 Siege of St. Augustine* (Fort Leavenworth: United States Army Command and General Staff College, 1997), 42n32.

29. Christopher Magra, *Poseidon's Curse: British Naval Impressment and Atlantic Origins of the American Revolution* (Cambridge: Cambridge University Press, 2016), 243–50.

30. "Vote relating to the Prov: Privateer, June 11 1740," Massachusetts State Archives Collection, Colonial Period, V. 62, p. 596, FamilySearch website.

31. Colonel William Stephens, June 4, 1740, in *Colonial Records of the State of Georgia*, vol. 4, *Stephens' Journal, 1737–40*, ed. in Allen D. Candler (Atlanta: Franklin Printing and Publishing, 1906), 586.

32. *South Carolina Gazette* (Charles Town), May 17–May 24, 1740, Accessible Archives.

33. Herson, *Joint Opportunity*, 36.

34. Legislature Minutes, South Carolina, November 8, 1739, in *The Journal of the Commons House of Assembly [JCHA], September 12, 1739–March 26, 1741*, ed. J. H. Easterby (Columbia: Historical Commission of South Carolina, 1952), 18–19. For more on this schooner, see the following sources: Assembly Minutes April 2, 1740, and July 18, 1740, 289–90 and 361–62, in *JCHA*; and William Bull to Col. Vanderdussen, July 9, 1740, in *Report of the Committee*, 103. Also see *An Impartial Account of the Late Expedition Against St. Augustine Under General Oglethorpe* (1742; repr., Gainesville: University Presses of Florida, 1978), 20–21; and *Report of the Committee*, 31.

35. Ivers, *British Drums*, 109–12; Baine, "General James Oglethorpe," 218–20.

36. Herson, *Joint Opportunity*, 22–25.

37. Trevor Reese, *Colonial Georgia: A Study in British Imperial Policy in the Eighteenth Century* (Athens: University of Georgia Press, 1963), 80.

38. Leach, *Roots of Conflict*, 47.

39. *Report of the Committee*, 143–45. Provincial critique did not mean that Anglo-Americans eschewed Royal Navy assistance. For instance, the colony's Commons House of Assembly still admitted a firm reliance on imperial protection. In a letter to King George II immediately following the withdrawal from St. Augustine, provincial legislators lamented the expedition's costs but thanked the monarch for the "Assistance of so many of your Majesty's Ships of War." See "The Humble Petition and Representation of the Council and Assembly of Your Majesty's Province of South Carolina Upon the Present State of the Said Province," July 26, 1740, in *JCHA, 1739–1741*.

40. Swanson, *Predators and Prizes*, 144–52.

41. Reese, *Colonial Georgia*, 80.

42. "Lieutenant Governor Bull to Duke of Newcastle and the affidavit of Thomas Lloyd and Robert Ford, Mariners, relating to their being taken by a Spanish sloop at St. Augustine, inclos'd," October 1741, Correspondence, TNA, Kew, CO 5/388, Adam Matthew, Colonial America Digital Database.

43. *JCHA 1739–1741*, August 16, 1740, 380–81.

44. Jack Schuler, *Calling Out Liberty: The Stono Slave Rebellion and the Universal Struggle for Human Rights* (Jackson: University Press of Mississippi, 2009), 69–72, ProQuest.

45. *JCHA 1739–1741*, September 10–16, 385–95, 425–26. Evidently the South Carolina government employed two galleys during the latter years of Queen Anne's War, so Bull's proposal was not without precedent. Thanks to Dr. Nic Butler for sharing this with me. See the October 24, 1707, entry in *Journal of the Commons House of Assembly of South Carolina, Transcript. Green Copy NO. 3, 1706–1711*, 307, MSS, South Carolina Departments of Archive and History.

It is noteworthy that Governor Tinker of the Bahamas made the same request for galleys but seems to have built them locally. See "Journal, July 1741: Volume 49," in *Journals of the Board of Trade and Plantations: Volume 7, January 1735—December 1741*, ed. K. H. Ledward (London: His Majesty's Stationery Office, 1930), 390–96, British History Online and the *South Carolina Gazette* (Charles Town), June 18–June 25, 1741, Accessible Archives.

46. "Letter from the Board of Trade to the Duke of Newcastle inclosing an extract from Mr. Bull, President of the Council in South Carolina, stating the danger that province is in from the row galleys which the Spaniards have at St. Augustine," Correspondence, TNA, Kew, CO 5/384, Adam Matthew, Colonial America Digital Database. Also, see Bull's letter to Newcastle, October 14, 1741, cited above, CO 5/388.

47. Charles C. Jones, Jr. *Collections of the Georgia Historical Society*, vol. 4, *The Dead Towns of Georgia* (Savannah: Morning News Steam Printing House, 1878), 98–99.

48. *Acts of the Privy Council of England, Colonial Series*, vol. 3, A.D. 1720–1745, ed. W. L. Grant, James Munro, and Sir Almeric W. Fitzroy (London: His Majesty's Stationery Office, 1910), 708–9.

49. See previous chapter for discussion on this policy.

50. Captains Log, HMS *Flamborough*, June 29, 1742, ADM 51/358, TNA.

51. Legislature Minutes, South Carolina, November 19–22, 1742, and December 1, 1742, in *Journal of the Commons House of Assembly September 14, 1742–January 27, 1744*, ed. J. H. Easterby (Columbia: South Carolina Archives Department, 1954), 40–43, 81–82.

Pennsylvania Gazette (Philadelphia), November 11, 1742, 2, Readex, America's Historical Newspapers.

52. Mark M. Smith, "African Dimensions," in *Documenting and Interpreting a Southern Slave Revolt* ed. John K. Thornton (Columbia: University of South Carolina Press, 2005), 79–80, JSTOR.

53. Barnett A. Elizas, *The Jews of South Carolina: From the Earliest Times to the Present Day* (Philadelphia: J. B. Lippincott, 1905), 28.

54. Huw David, *Trade, Politics, and Revolution: South Carolina and Britain's Atlantic Commerce, 1730–1790* (Columbia: University of South Carolina Press, 2018), 43–62, JSTOR.

55. *Proceedings and Debates of the British Parliaments Respecting North America*, vol. 5, 1739–1754, ed. Leo Francis Stock (Washington, DC: Carnegie Institution of Washington, 1941), 126–29; Chapin, *Privateering in King George's War*, 32–34.

56. *Proceedings and Debates of the British Parliaments*, 130–31.

57. Richmond, *Navy in the War of 1739–48*, 270–73.

58. *The Scots Magazine . . . For the Year MCCCXLII*, vol. 4 (Edinburgh: Sands, Brymer, Murray and Cochran), 285.

59. W. E. May, "Capt. Charles Hardy on the Carolina Station, 1742–1744," *South Carolina Historical Magazine* 70, no. 1 (1969): 2, JSTOR. Whatever the tone change in these instructions, many provincial authorities in Charles Town did not consider Hardy's arrival to be an improvement upon Fanshawe, with some South Carolina elites accusing Hardy of being as inactive as his predecessor. See Swanson, *Predators and Prizes*, 173–74.

60. Swanson, *Predators and Prizes*, 173–74.

61. Ivers, *British Drums*, 144–47.

62. General Oglethorpe to Harman Verelst, December 7, 1741, in *Trustees for Establishing the Colony of Georgia in America: "Letters from Georgia, v. 14206, 1741 June–1742 December,"* Digital Library of Georgia, 49–51.

63. "A list of the military strength of Carolina and Georgia," List, TNA, Kew, CO 5/655, Adam Matthew, Colonial America Digital Database.

64. W. E. May, "Captain Frankland's *Rose*," *American Neptune* 26 (1966): 43–47; and "Capt. Charles Hardy," 6.

65. Council Minutes, South Carolina, July 19, 1742, in "Council minutes relating to Captain Charles Hardy of HMS Rye," Minutes, Correspondence, Transcript, TNA, Kew, CO 5/369, Adam Matthew, Colonial America Digital Database.

66. "Charles II, 1661: An Act for the Establishing Articles and Orders for the regulateing and better Government of His Majesties Navies Ships of Warr & Forces by Sea," in *Statutes of the Realm*, vol. 5, 1628–80, ed. John Raithby, 311–14 (s.l: Great Britain Record Commission, 1819), British History Online.

67. Ivers, *British Drums*, 153–60; Margaret Davis Cate, "Fort Frederica and the Battle of Bloody Marsh" *Georgia Historical Quarterly* 27, No. 2 (June 1943): 142–43, JSTOR.

68. Ivers, *British Drums*, 159–72.

69. Colonel Fenwicke to Captain Hardy, July 30, 1742, in "Council minutes relating to Captain Charles Hardy of HMS Rye," Minutes, Correspondence, Transcript, TNA, Kew, CO 5/369, Adam Matthew, Colonial America Digital Database.

70. May, "Captain Frankland's *Rose*," 46–47.

71. "Charles-Town, (South-Carolina) August 30," *American Weekly Mercury* (Philadelphia), October 14, 1742, 2, Readex, America's Historical Newspapers.

72. Ivers, *British Drums*, 183.

73. Amos Aschbach Ettinger, *James Edward Oglethorpe, Imperial Idealist* (Oxford: Clarendon Press, 1936), 252–53.

74. "An Account of Extraordinary Services Incurred in Georgia, for the Preservation and Defence of his Majesty's Dominions on the Continent of North America [. . .]," in *Journals of the House of Commons*, vol. 24 (repr., House of Commons, 1803), 615.

75. Ivers, *British Drums*, 185.

76. For more on the political machinations that led to this conflict, see Jeremy Black, *America or Europe?: British Foreign Policy, 1739–63* (London: Taylor & Francis Group, 1997).

77. Swanson, *Predators and Prizes*, 30–38.

78. "An Act for the Better Encouragement of Seamen in his Majesty's Service, and Privateers, to annoy the Enemy," in *The Statutes at Large, From the 15th to the 20th Year of George II*, vol. 18, ed. Danby Pickering (Cambridge: Joseph Bentham, 1765), 250–53.

79. For an example, see "Charles-Town, South-Carolina, July 9," *American Weekly Mercury* (Philadelphia), August 20, 1741, 2, Readex, America's Historical Newspapers.

80. George Rawlyk, *Yankees at Louisbourg* (Orono: University of Maine Press, 1967), 1–5.

81. Rawlyk, *Yankees*, 6–10.

82. Paullin, *Colonial Army and Navy*, 48–55; and Harold E. Selesky, *War and Society in Colonial Connecticut* (New Haven: Yale University Press, 1990), 73–74.

83. Qtd. in Chapin, *Privateering*, 100–101.

84. Swanson, *Predators and Prizes*, 84–87.

85. *Pennsylvania Gazette* (Philadelphia) July 12, 1744, 1, Readex, America's Historical Newspapers.

86. "An Act for the more effectual securing and encouraging the Trade of His Majesty's British Subjects to America, and for the Encouragement of Seamen to enter into His Majesty's Service," in *The Statutes at Large of England and of Great-Britain: From Magna Carta to the Union of the Kingdoms of Great Britain and Ireland*, vol. 9, ed. John Raithby, 662–68 (London: George Eyre and Andrew Strahan, 1811).

87. *Journal of the Commissioners for Trade and Plantations, From January 1741-2 to December 1749* (London: His Majesty's Stationery Office, 1931), 132.

88. Francis Fane to the Board of Trade, November 7, 1744, in "Documents concerning colonial finances and war with France," Correspondence, Legislation, Order, Report, TNA, Kew, CO 5/884, Adam Matthew, Colonial America Digital Database.

89. *Acts and Resolves, Public and Private*, 216–18.

90. Commodore Peter Warren to George Clinton, July 6, 1744, and Thomas Corbett to Commodore Peter Warren, March 18, 1744/5, in *Royal Navy and North America*, ed. Julian Gwyn, 33, 64–65. Also see Douglas, *Nova Scotia and the Royal Navy*, 46–47.

91. Warren to Anson, April 2, 1745, 70–74.

92. Douglas, *Nova Scotia and the Royal Navy*, 90.

93. Rawlyk, *Yankees at Louisbourg*, 31–54, 78–79.

94. Louis Effingham de Forest, ed. "Appendix I: The Fleet," in *Louisbourg Journals 1745* (New York: Society of Colonial Wars in New York, 1932), 181–83; and Howard Chapin, Howard Chapin, "New England Vessels in the Expedition against Louisbourg, 1745," *New England Historical and Genealogical Register* 76 (January 1922): 59–60.

95. "Boston, February 25," *Pennsylvania Gazette* (Philadelphia), March 26, 1745, 2, Readex, America's Historical Newspapers.

96. Governor William Shirley to the Duke of Newcastle, March 27, 1745, in *Correspondence of William Shirley, Governor of Massachusetts and Military Commander in America, 1731–1760*, ed. Charles Henry Lincoln (New York: Macmillan, 1912), 196–98.

97. Douglas, in *Nova Scotia and the Royal Navy*, makes the case that the Admiralty did not suddenly develop interest in military operations in Canada, but its interest in more decisive action in northern waters grew over time (38–68). For a more concise view of the creation of the North American Squadron, see Julian Gwyn, *Frigates and Foremasts: The North American Squadron in Nova Scotia Waters 1745–1815* (Vancouver: University of British Columbia Press, 2003), 9.

98. Douglas, *Nova Scotia and the Royal Navy*, 62–68.

99. "Legislative Acts/Legal Proceedings," *Boston Gazette*, July 23, 1745, 1, Readex, America's Historical Newspapers.

100. Chapin, *Privateering in King George's War*, 90–92.

101. Bedford to Newcastle, March 24, 1745/6, in *Royal Navy and North America*, ed. Julian Gwyn, 223–26 and Douglas, *Nova Scotia and the Royal Navy*, 198.

102. Nagel, *Empire and Interest*, 502–3; and Glyndwr Williams, *The British Atlantic Empire before the American Revolution* (London: Taylor & Francis Group, 1980), 78–80. For an example of the specific provincial naval costs that Parliament reimbursed, see *Journals of the House of Commons* [. . .], vol. 25 (London: House of Commons, 1803), 1042–43.

103. Chapin, *The Tartar*, 23n60.

104. Swanson, *Predators and Prizes*, 38–41.

105. Chapin, *Privateering in King George's War*, 47–52.

106. "Richardson & others V. Ship Two Friends & Cargo, Decree," in Suffolk County (MA) Court Files, 1629–1797, v. 384, Case 61447, FamilySearch website.

107. "An Act for the Better Encouragement of Seamen," 261.

108. "Richards & others V. Ship Two Friends & Cargo, Decree."

109. Nathaniel Sparhawk to William Pepperell, January 14, 1745/6, in *Collections of the Massachusetts Historical Society, Sixth Series*, vol. 10 (Boston: Massachusetts Historical Society, 1899), 428–29. It is important to note that these tensions were not relegated to maritime disputes. For more information on controversies between New England troops and their Redcoat compatriots during the Siege of Louisbourg, see Leach, *Roots of Conflict*, 68–72.

110. J. Revell Carr, *Seeds of Discontent: The Deep Roots of the American Revolution, 1650–1750* (New York: Bloomsbury, 2008), Kindle eBook Edition, Loc. 4476.

111. Chapin, *The Tartar*, 23n60.

112. "Notes of Arguments Advanced by Dr. Pinfold and Mr. Yorke," May 17, 1749, in

Prize Appeals, 1751–1758, vol. 2, fol. 256, ed. Sir George Lee, New York Public Library Digital Collections.

113. "Notre Dame de Deliverance . . . The Case of the Three Raspondents," May 3, 1750, in *Prize Appeals, 1736–1751*, vol. 1, fol. 65, ed. Sir George Lee. New York Public Library Digital Collections.

114. "The Case of his Majesty's Ships Chester and Sunderland, the actual and sole Captors of the Prize" July 5, 1750, in *Prize Appeals, 1736–1751*, vol. 1, fol. 70–73, ed. Sir George Lee.

115. Untitled written notes below title page on "The Appellants Case," in *Prize Appeals, 1736–1751*, vol. 1, fol. 75–76, ed. Sir George Lee; and *Gentleman's Magazine and Historical Chronicle*, vol. 20, *For the Year MDCCL* (London: Edward Cave, 1750), 328.

116. Carr, *Seeds of Discontent*, loc. 5047–5132.

117. For the concept of provincial navies and privateers "augmenting" the Royal Navy, see Richmond, *Navy in the War of 1739–48*, 274; and Swanson, *Predators and Prizes*, 3.

118. Henry Laurens to James Crokatt, December 28, 1747, in *The Papers of Henry Laurens*, vol. 1, *Sept. 11, 1746–Oct. 31, 1755*, ed. Philip M. Hamer, George C. Rogers, Maude E. Lyles (Columbia: University of South Carolina Press, 1968), 96.

119. "Extract from a Letter From Antigua . . ." qtd. in *The History of the Island of Antigua, One of the Leeward Caribees in the West Indies [. . .]*, vol. I (London: Mitchell and Hughes, 1891), cv.

120. Harding, *Seapower*, pp. 195–98, "The War in the West Indies," in *The Seven Years' War: Global Views* eds. Mark Danley and Patrick Speelman (Leiden: BRILL, 2012), 293–306, ProQuest.

121. Richmond, *Navy in the War of 1739–48*, 242-244.

122. Magra, *Poseidon's Curse*, 250–51.

123. Jack Tager, *Boston Riots: Three Centuries of Social Violence* (Boston: Northeastern University, 2001), 61–64.

124. Magra, *Poseidon's Curse*, 284–86.

125. William Shirley to the Lords of Trade, December 1, 1747, in *Correspondence of William Shirley*, 414.

126. Magra, *Poseidon's Curse*, 291–93.

127. Ivers, *British Drums*, 195–202; and James M. Johnson, *Militiamen, Rangers, and Redcoats: The Military in Georgia, 1754–1776* (Macon: Mercer University Press, 1992), Google Play eBook, 12.

128. Sirmans, *Colonial South Carolina*, 276.

129. Douglas, *Nova Scotia and the Royal Navy*, 151.

130. Douglas, "Sea Militia," 22–37.

CHAPTER 5

1. A. J. B. Johnson, *Endgame 1758: The Promise, the Glory, and the Despair of Louisbourg's Last Decade* (Lincoln: University of Nebraska Press, 2008), 32–45, ProQuest. Also see Grenier, *Far Reaches*, 141–71; and Douglass, *Nova Scotia and the Royal Navy*, 176–211.

2. Johnson, *Endgame*, 47–60; and Douglas, *Nova Scotia and the Royal Navy*, 189–91.

3. John R. Dull, *The French Navy and the Seven Years War* (Lincoln: University of Nebraska Press, 2005), 15–16.

4. Rodger, *Command of the Ocean*, 259–63.

5. Daniel Baugh, *The Global Seven Years War, 1754–1763* (Abingdon: Routledge, 2011), loc. 6165–50, Kindle eBook edition.

6. Fred Anderson, *Crucible of War: The Seven Years' War and the Fate of Empire in British North America, 1754–1766* (New York: Alfred A. Knopf, 2000), 72–79.

7. Andrew D. M. Beaumont, *Colonial America and the Earl of Halifax* (Oxford: Oxford University Press, 2015), 135–43.

8. See Benjamin Franklin, "Plain Truth, 17 November 1747," and "From Benjamin Franklin to James Alexander and Cadwallader Colden with Short Hints towards a Scheme for Uniting the Northern Colonies, 8 June 1754," and "Reasons and Motives for the Albany Plan of Union, [July 1754]," Founders Online/National Archives.

9. "Mr. Pownall's Consideration Towards a General Plan of Measures for the Colonies," in Matthew S. Quay, *Pennsylvania Archives, Second Series*, vol. 6 (Harrisburg: Lane S. Hart, 1877), 201. For a detailed analysis of Franklin's role in the Albany Congress as well as other delegates' proposals (and naval considerations), see Robert Clifford Newbold, *The Albany Congress and Plan of Union of 1754* (New York: Vantage Press, 1955), 74–131.

10. Beaumont, *Colonial America and the Earl of Halifax*, 149–50.

11. "Representation to the King on the Proceedings of the Congress at Albany," in *Documents Relative to the Colonial History of the State of New-York*, vol. 6, ed. E. B. O'Callaghan, 903–18 (Albany: Weed, Parsons and Company, 1855).

12. Anderson, *Crucible of War*, 85–90.

13. "At a Council held at the Camp at Alexandria [. . .]," in *Minutes of the Provincial Council of Pennsylvania*, vol. 6 (Harrisburg: Theo Fenn, 1851), 366–68.

14. Malcolm Macleod, *French and British Strategy In the Lake Ontario Theatre of Operations, 1754–1760*, unpublished graduate thesis, University of Ottawa, 1973, 258–72.

15. Robert Malcomson, "Not Very Much Celebrated: The Evolution and Nature of the Provincial Marine, 1755–1813," *Northern Mariner* 11, no. 1 (January 2001): 25–29; and Malcomson, *Warships of the Great Lakes, 1754–1834* (Annapolis: Naval Institute Press, 2001), 8–11. Special thanks to Sam Ewell for helping me to find a copy of this monograph.

16. Joseph F. Meany, "Batteau and 'Battoe Men': An American Colonial Response to the Problems of Logistics in Mountain Warfare," manuscript published digitally by New York State Military Museum, 2–10.

17. William C. Godfrey, *Pursuit of Profit and Preferment in Colonial North America: John Bradstreet's Quest* (Waterloo: Wilfrid Laurier University Press, 1982), 107–10, Google Play eBook edition.

18. Daniel R. Bazan, *For Want of Sloops, Water Casks, and Rum: The Difficulty of Logistics in the Canadian Theater of the Seven Years War*, unpublished master's thesis, Liberty University, 2013, 81–91.

19. Malcomson, *Warships of the Great Lakes*, 19.

20. Philip Chadwick Foster Smith, "King George, the Massachusetts Province Ship,

1757–1763: A Survey," in *Seafaring in Colonial Massachusetts, Publications of the Colonial Society of Massachusetts*, vol. 52, ed. Frederick S. Allis Jr., 175–85 (Boston: Colonial Society of Massachusetts, 1980).

21. For more on ministerial politics during this period, see Rodger, *Command of the Ocean*, 266–68. Governor Thomas Pownall to Pitt, August 16, 1757, in *Correspondence of William Pitt When Secretary of State*, vol. 1, ed. Gertrude Selwyn Kimball (New York: Macmillan, 1906), 98.

22. Francis Bernard to John Pownall, October 20, 1762, in *The Papers of Francis Bernard, Governor of Colonial Massachusetts*, vol. 1, *1759–1763*, ed. Colin Nicolson, 278–79 (Boston: Colonial Society of Massachusetts, 2007).

23. Rodger, *Command of the Ocean*, 277; Thomas M. Truxes, "The Breakdown of Borders: Commerce Raiding during the Seven Years' War, 1756–1763," in *Commerce Raiding: Historical Case Studies, 1755–2009*, ed. Bruce A. Elleman and S. C. M. Paine (New Port: Naval War College, 2013), 16–18.

24. *The Public Records of the Colony of Connecticut, From May, 1757, to March, 1762, Inclusive*, ed. Charles J. Hoadly (Hartford: Case, Lockwood, & Barnard, 1880), 9, 62, 109.

25. *Records of the Colony of Rhode Island and Providence Plantations*, vol. 6, *1757 to 1769*, ed. John Russell Bartlett (Providence: Knowles, Anthony, 1861), 22, 27, 38, 66, 162; and Samuel Greene Arnold, *History of the State of Rhode Island and Providence Plantations*, vol. 2, *1700–1790* (New York: D. Appleton, 1860), 204n1.

26. *Pennsylvania Gazette* (Philadelphia), July 28, 1757, 3, Readex, America's Historical Newspapers.

27. Legislature Minutes, Pennsylvania, November 24, 1758, in *Pennsylvania Archives, Eighth Series*, vol. 6, *October 14, 1756–January 3, 1764*, ed. Charles F. Hoban (Philadelphia: Pennsylvania State Library, 1935), 4899; and *New-York Mercury*, January 1, 1759, 1, Readex, America's Historical Newspapers.

28. William H. Egle, *History of the Commonwealth of Pennsylvania, Civil, Political, and Military, from Its Earliest Settlement to the Present Time* (Philadelphia: E. M. Gardner, 1883), 1027.

29. *Boston News-Letter*, September 8, 1757, 1, Readex, America's Historical Newspapers.

30. *Pennsylvania Gazette* (Philadelphia), August 4, 1757, 2, Readex, America's Historical Newspapers.

31. *New Hampshire Gazette* (Portsmouth, New Hampshire), September 1, 1758, Readex, America's Historical Newspapers.

32. Beaumont, *Colonial America and the Earl of Halifax*, 198–201.

33. Pares, *War and Trade*, 296–98.

34. Michael J. Jarvis, *In the Eye of All Trade: Bermuda, Bermudians, and the Maritime Atlantic World, 1680–1783* (Chapel Hill: Omohundro Institute of Early American History & Culture, 2010), 245.

35. "His Honour the Lieutenant Governor's Speech at the Opening of a Sessions of the Council & Assembly of Jamaica on Tuesday the 27th day of September 1757," TNA, Kew, CO 137/30, 46.

36. David Starkey, *British Privateering Enterprise in the Eighteenth Century* (Exeter: University of Exeter Press, 1990), 137–38, 161–79.

37. Gary B. Nash, *The Urban Crucible: The Northern Seaports and the Origins of the American Revolution* (Cambridge: Harvard University Press, 1986), 147–50. Also see Michael Watson, "Judge Lewis Morris, the New York Vice-Admiralty Court, and Colonial Privateering, 1739–1762," *New York History*, 78, no. 2 (April 1997): 145, JSTOR.

38. Douglas, *Nova Scotia and the Royal Navy*, 208–11.

39. Martin Robson, *A History of the Royal Navy : The Seven Years War* (London: I. B. Tauris, 2016), 15–26, EBSCOhost. Also see Rodger, *Command of the Ocean*, 259.

40. Rodger, *Command of the Ocean*, 268–70.

41. Robson, *History of the Royal Navy*, 30–42.

42. Robson, *History of the Royal Navy*, 83–85.

43. Rodger, *Command of the Ocean*, 274–75.

44. Gould, *Persistence of Empire*, 38–65.

45. Rodger, *Command of the Ocean*, 276–77; and *Pennsylvania Gazette* (Philadelphia), July 27, 1758, 2, Readex, America's Historical Newspapers.

46. Harding, "The War in the West Indies," 316–22.

47. *South Carolina Gazette* (Charles Town), May 22, 1762, Accessible Archives.

48. Harding, *Seapower and Naval Warfare*, 206–11. For more on the Royal Navy's improvements in supplies, see Buchet, Higgie, and Duffy, *British Navy, Economy and Society*.

49. Nagel, *Empire and Interest*, 502–8.

50. *Records of the Colony of Rhode Island*, 209.

51. Anderson, *Crucible of War*, 316–24.

52. Leach, *Roots of Conflict*, 156–57.

53. Alan Rogers, *Empire and Liberty: American Resistance to British Authority, 1755–1763* (Berkeley: University of California Press, 1974), 46–48.

54. Jesse Lemisch, *Jack Tar vs. John Bull: The Role of New York's Seamen in Precipitating the Revolution* (New York: Routledge, 1997), loc. 516–32, Kindle eBook edition.

55. Lemisch, *Jack Tar*, loc. 801.

56. Leach, *Roots of Conflict*, 158–62.

57. Gould, *Among the Powers*, 83–93.

CHAPTER 6

1. Nash, *Urban Crucible*, 157–59, 170–75.

2. Caleb Hopkins Snow, *A History of Boston: The Metropolis of Massachusetts, From Its Origin to the Present Period* (Boston: A Bowen, 1828), 251–52.

3. James Otis, *A Vindication of the Conduct of the House of Representatives of the Province of the Massachusetts-Bay* [. . .] (Boston: Edes & Gill, 1762), 11–15.

4. Otis, *Vindication*, 33.

5. Kinkel, *Disciplining the Empire*, 164–65.

6. Johnson, *Militiamen, Rangers, and Redcoats*, 56–58.

7. *Newport (RI) Mercury*, February 10, 1766, Readex, America's Historical Newspapers.

8. "Extracts from the statement of the services and expenses of the Province of Massachusetts, made by a Committee of the Council and House of Representatives,

chosen for the purpose in October, 1764, and sent to the colony's agent in England, to furnish arguments why the colony should not be taxed, &c.," in *Speeches of the Governors of Massachusetts, From 1765 to 1775 [. . .]*, ed. Alden Bradford (Boston: Russell and Gardner, 1818), 25–27.

9. Stephen Hopkins, *The Rights of the Colonies Examined* (Providence: William Goddard, 1765), 20–21, Evans Early Imprint Collection; and Raymond G. O'Connor, *Origins of the American Navy: Sea Power in the Colonies and the New Nation* (Lanham: University Press of America, 1994), 15.

10. Chapin, *Tartar*, 51–2.

11. Stout, *Royal Navy*, 27–30.

12. Stout, *Royal Navy*, 59, 128–29.

13. Michael R. Derderian, "This Licentious Republic: Maritime Skirmishes in Narragansett Bay 1763–1769," *Journal of the American Revolution*, Electronic Journal, October 2, 2017; and Stout, *Royal Navy*, 69–70.

14. Stout, *Royal Navy*, 66–68; and Lt. Thomas Hill, "Remarks on board His Majesty's schooner, the St. John, in Newport harbor, Rhode Island," in *Records of the Colony of Rhode Island and Providence Plantations in New England*, vol. 6, ed. John Russell Bartlett, 428–29 (Providence: Knowles, Anthony, State Printers, 1861).

15. Derderian, "This Licentious Republic"; and Stout, *Royal Navy*, 118–22.

16. Smith, "King George," 179–80.

17. Stout, *Royal Navy*, 141–42.

18. Admiral Montagu to Governor Wanton, April 8, 1772, and Governor Wanton to Admiral Montagu, May 8, 1772, in *Records of the Colony of Rhode Island and Providence Plantations in New England*, ed. John Bartlett, vol. 7, 62–65.

19. Ephraim Bowen, "Narrative of the Capture and Burning of the British Schooner Gaspee," in *Records of the Colony of Rhode Island and Providence Plantations in New England*, ed. John Bartlett, vol. 7, 68–72. The narrative that a maritime mob assembled by a drum beat in town is echoed by Admiral Montagu himself in a letter to the governor (Montagu to Wanton, 88–89).

20. *Boston News-Letter*, April 24, 1704, 2, Readex, America's Historical Newspapers; and Governor and Company of Rhode Island to the Council of Trade and Plantations, September 14, 1706, in *CSP*, vol. 23, *1706–1708*, ed. Cecil Headlam, 213–30 (London: His Majesty's Stationery Office, 1916), British History Online.

21. Chapin, *Privateer Ships*, 69–70; and John Osborne Austin, *The Genealogical Dictionary of Rhode Island* (Albany: Joel Munsell's Sons, 1887), 215–18.

22. Pauline Maier, *From Resistance to Revolution: Colonial Radicals and the Development of American Opposition to Britain, 1765–1776* (New York: Alfred A. Knopf, 1973), 11–16.

23. For example, see Paul A. Gilje, *Liberty on the Waterfront: American Maritime Culture in the Age of Revolution* (Philadelphia: University of Pennsylvania Press, 2007), 99–100. Marcus Rediker and Peter Linebaugh, *The Many-Headed Hydra: The Hidden History of the Revolutionary Atlantic* (Boston: Beacon Press, 2000), 220–21.

24. Nash, *Urban Crucible*, 147–55.

25. Lemisch, *Jack Tar*, loc.1432–46, 2064–3502.

26. See Magra, *Poseidon's Curse*; Gilje, *Liberty on the Waterfront*, 99–105.

27. Magra, *Poseidon's Curse*, 294–312.

28. Stout, *Royal Navy*, 144.

29. John Adams to E. Gerry, February 11, 1813, in *Extracts Relating to the Origin of the American Navy*, ed. Henry E. Waite (Boston: New England Historic Genealogical Society, 1890), 3.

30. Sam Willis, *The Struggle for Sea Power: A Naval History of American Independence* (New York W. W. Norton, 2016), 40–44, Kindle eBook edition. For an early example of a historian placing the navy's birth in New England, see Charles Oscar Paulin, *The Navy of the American Revolution: Its Administration, Its Policy and its Achievements*, published diss., University of Chicago, 1906, 32–36.

31. Christopher Magra, *The Fisherman's Cause: Atlantic Commerce and Maritime Dimensions of the American Revolution* (Cambridge: Cambridge University Press, 2009), 177–78, Kindle eBook edition.

32. Paullin, *Navy*, 34–36.

33. George C. Daughn, *If By Sea: The Forging of the American Navy: From the American Revolution to the War of 1812* (New York: Basic Books, 2008), 35–42, Kindle eBook edition.

34. "Stephen Hopkins, 1707–1785," United States House of Representatives website.

35. Paullin, *Navy*, 315–18.

36. "An act for the encouraging the fitting out of armed vessels to defend the sea coast of America, and for erecting a court to try and condemn all vessels that shall be found infesting the same," November 13, 1775, qtd. in James T. Austin, *The Life of Elbridge Gerry* (Boston: Wells and Lilly, 1828), 524; and Paullin, *Navy*, 321–23.

37. D. E. Huger Smith, "Commodore Alexander Gillon and the Frigate South Carolina," *South Carolina Historical and Genealogical Magazine* 9, no. 4 (1908): 194.

38. James A. Lewis, *Neptune's Militia: The Frigate South Carolina during the American Revolution* (Kent: Kent University Press, 1999), 15–18, EBSCOhost.

39. Smith, "Commodore Alexander Gillon," 195–200.

40. Lewis, *Neptune's Militia*, 6–18.

41. Smith, "Commodore Alexander Gillon," 215–17.

42. Smith, "King George," 176–85.

43. Johnson, *Militiamen, Rangers, and Redcoats*, 12. For life expectancy of eighteenth-century vessels, see Fernand Braudel, *Civilization and Capitalism, 15th–18th Century*, vol. 2, *The Wheels of Commerce*, trans. Sian Reynolds (Berkeley: University of California Press), 369.

44. Maya Jasanoff, *Liberty's Exiles: American Loyalists in the Revolutionary World* (New York: Vintage Books, 2011), 42.

45. Joseph Habersham to William Henry Drayton, "Letter to William Henry Drayton Concerning Recruiting of Sailors for the Ship Prosper [. . .]," c. 1776, 258–59, Robert W. Gibbes Collection of Revolutionary War Manuscripts (Series 213089 Box: 0004), South Carolina Department of Archives and History website.

46. Gordon Burns Smith, *Morningstars of Liberty: The Revolutionary War in Georgia, 1775–1783* (Milledgeville, GA: Boyd Pub., 2006), 334.

47. Legislature Minutes, Georgia, February 10, 1778, in *The Revolutionary Records*

of the State of Georgia, vol. 2, ed. Allen D. Candler (Atlanta: Franklin-Turner, 1908), 43–44.

48. J. G. Braddock, "The Plight of a Georgia Loyalist: William Lyford, Jr.," *Georgia Historical Quarterly* 91, no. 3 (2007): 247–65.

49. George E. Buker, and Richard Apley Martin. "Governor Tonyn's Brown-Water Navy: East Florida during the American Revolution, 1775–1778," *Florida Historical Quarterly* 58, no. 1 (1979): 58–71.

Conclusion

1. Chapin, "New England Vessels," 59–60.

Bibliography

Description of Manuscript Collections Consulted

Most of the primary sources consulted in this book come from the Adam Matthew Colonial America digital database, which includes scans from the Colonial Office Papers at The National Archives in Kew, England, and are often under the heading "CO 5." Other documents that could not be digitized such as Royal Navy ship logs and Caribbean legislative records were consulted in person on trips to the archives. References to documents from the archives will be referred to as TNA. They will also be listed under different CO (Colonial Office) or ADM (Admiralty) headings.

Many Massachusetts archives documents also come from digitized copies on FamilySearch. Another major source has been the Calendar of State Papers series, which summarizes many of the CO papers up to the year 1739, and frequently directly quotes from the original sources. The calendar will be referred to below, as in the footnotes, as *CSP*. Finally, all of the colonial newspapers consulted in this book came from digital databases provided by the College of Charleston and the University of New Hampshire, including *Accessible Archives* and *America's Historical Newspapers*.

Primary Sources

Correspondence/Petitions/Speeches/Laws/Legal Briefs/Military Orders

"An Account of Extraordinary Services Incurred in Georgia, for the Preservation and Defence of His Majesty's Dominions on the Continent of North America [. . .]." In *Journals of the House of Commons*, vol. 24. Repr. ed., House of Commons, 1803.

"Account of moneys disbursed and paid for a new sloop built by John Cooke for their majesties service." May 8, 1691." Financial Document, TNA, Kew, CO 5/856. Adam Matthew, Colonial America Database.

"An Account of the Invasion of South Carolina by the French & Spaniards in the Month of August 1706." In *Records in the British Public Record Office Relating to South Carolina 1701–1710*, edited by A. S. Salley. Columbia: Historical Commission of the State of South Carolina, 1947.

"An Act for Appointing Two Scout Canoes." In *South Carolina Statutes at Large*, vol. 2, edited by Thomas Cooper. Columbia, SC: A. S. Johnston, 1837.

"An act for carrying on several expeditions against the Indian and other enemies and for defraying the charge thereof." September 30, 1727. Legislation, TNA, Kew, CO 5/387. Adam Matthew, Colonial America Digital Database.

"An Act for the Better Encouragement of Seamen in His Majesty's Service, and Privateers, to annoy the Enemy." In *The Statutes at Large, From the 15th to the 20th Year of George II*, vol. 18, edited by Danby Pickering. Cambridge: Joseph Bentham, 1765.

"An Act for the Better Security of That Parte of the Province of Carolina . . . Against Any Hostile Invasions and Attempts by Sea or Land, Which the Neighbouring Spaniard or Other Enemy May Make Upon the Same." In *The Statutes at Large of South Carolina*, vol. 2, from 1682–1716, edited by Thomas Cooper. Columbia, SC: A. S. Johnston, 1837.

"An act for the encouraging the fitting out of armed vessels to defend the sea coast of America, and for erecting a court to try and condemn all vessels that shall be found infesting the same." November 13, 1775. Qtd. in James T. Austin, *The Life of Elbridge Gerry*. Boston: Wells and Lilly, 1828.

"An Act for the more effectual securing and encouraging the Trade of His Majesty's British Subjects to America, and for the Encouragement of Seamen to enter into His Majesty's Service." In *The Statutes at Large of England and of Great-Britain: From Magna Carta to the Union of the Kingdoms of Great Britain and Ireland*, vol. 9, edited by John Raithby. London: George Eyre and Andrew Strahan, 1811.

Acts of the Privy Council of England, Colonial Series. Vol. 3, A.D. 1720–1745. Edited by W. L. Grant, James Munro, and Sir Almeric W. Fitzroy. London: His Majesty's Stationery Office, 1910.

Adams, John, to E. Gerry. February 11, 1813. In *Extracts Relating to the Origin of the American Navy*, edited by Henry E. Waite. Boston: New England Historic Genealogical Society, 1890.

"An Agreement made by order of His Excellency Francis Nicholson . . . By Coll: John Barnwell . . . With Jonathan Collings Commander of the Sloop Jonathan and Sarah [. . .]." June 9, 1721. In "Letters and papers relating to the landing of his Majesty's Independent Company now in south Carolina, and likewise concerning Colonel Barnwell's going to Altamaha river in order to build a fort there." Correspondence, Submission, Warrant, TNA, Kew, CO 5/358. Adam Matthew, Colonial America Digital Database.

Andros, Edmund, to John Cooke. August 6, 1687. Massachusetts State Archives, vol. 127. FamilySearch website.

"An Answer to the Queries sent by the Honble the Lords Commissioners of trade and plantations relating to the State of South Carolina." William R. Coe Collection, 1699–1741, South Carolina Historical Society.

"At a Council held at the Camp at Alexandria [. . .]." In *Minutes of the Provincial Council of Pennsylvania*, vol. 6. Harrisburg: Theo Fenn, 1851.

"Attacks by the Dutch on the Virginia Fleet in Hampton Roads in 1667." *Virginia Magazine of History and Biography* 4, no. 3 (January 1897): 229–45.

"At the Court at Whitehall, the 26th: Aprill 1690." In "New England patents and grants, 1690," 229–30. Submission, Order, TNA, Kew, CO 5/905. Adam Matthew, Colonial America Database.

Bedford to Newcastle. March 24, 1745/6. In *Royal Navy and North America*, edited by Julian Gwyn, 223–26. Navy Records Society, 1973.

Bernard, Governor Francis, to John Pownall. October 20, 1762. In *The Papers of Francis Bernard, Governor of Colonial Massachusetts*, vol. 1, *1759–1763*, edited by Colin Nicolson. Boston: Colonial Society of Massachusetts, 2007.

Bowen, Ephraim. "Narrative of the Capture and Burning of the British Schooner Gaspee." In *Records of the Colony of Rhode Island and Providence Plantations*, edited by John Bartlett, vol. 7, 68–72.

Burchett, J., to the Lords of the Treasury. December 19, 1723 (No. 49). In "Volume 248: July 3–December 31, 1724," in *Calendar of Treasury Papers*, vol. 6, *1720–1728*, edited by Joseph Reddington, 276–98. London: Her Majesty's Stationery Office, 1889. British History Online.

"Byfield's letter to Dudley about assemblies." June 12, 1694. Correspondence, TNA, Kew, CO 5/858. Adam Matthew, Colonial America Database.

"Charles II, 1661: An Act for the Establishing Articles and Orders for the regulateing and better Government of His Majesties Navies Ships of Warr & Forces by Sea." In *Statutes of the Realm*, vol. 5, *1628–80*, edited by John Raithby. Great Britain Record Commission, 1819, 311–14. British History Online.

"The Charter of Massachusetts Bay: 1629." Yale University, Avalon Project website.

The Charters and General Laws of the Colony and Province of Massachusetts Bay. Boston: T. B. Wait, 1814.

Codrington, Governor Christopher, to the Lords of Trade and Plantations. November 1689. TNA, Kew, CO 153/4.

Cooke, Elisha, and Thomas Oakes. "An Answer to Sr: Edmond Andros's Acco: of the Forces Raised in New England for Defence of the Country against the Indians. . . ." 30 May 1690." In "New England patents and grants, 1690." Submission, Order, TNA, Kew, CO 5/905. Adam Matthew, Colonial America Database.

"Copy of Captain Montjoys commission and instructions." Commission, Order, TNA, Kew, CO 5/387.

Corbett, Thomas, to Commodore Peter Warren. March 18, 1744/5. In *The Royal Navy and North America: The Warren Papers, 1736–1752*, edited by Julian Gwyn. Navy Records Society, 1973.

"Cost of Preparing West Indies Expedition." Circular. Sent to Connecticut, Maryland, Massachusetts, New Hampshire, New Jersey, New York, North Carolina, Rhode Island, and [Virginia?], April 2, 1740. In *Royal Instructions to British Colonial Governors, 1670–1776*, vol. 3, edited by Leonard Labaree. New York: D. Appleton, 1935.

Council of Trade and Plantations to Mr. Secretary Craggs. December 14, 1720 (No. 322). In *CSP*, vol. 32, *1720–1721*, edited by Cecil Headlam, 212–28. London: His Majesty's Stationery Office, 1933. British History Online.

Council of Trade and Plantations to the Lords Justice. September 23, 1720. In *CSP*, vol.

32, 1720–1721, edited by Cecil Headlam, 144–65. London: His Majesty's Stationery Office, 1933. British History Online.

Cumings, Archibald, to William Popple. June 20, 1722. In "Letters from Samuel Shute and Archibald Cumings to the Board of Trade, enclosing Council Minutes." Correspondence, Minutes, TNA, Kew, CO 5/868. Adam Matthew, Colonial America Database.

Delafaye, Charles, to Governor Henry Worsley. August 1723. In *CSP*, vol. 33, 1722–1723, edited by Cecil Headlam, 221–38. London: His Majesty's Stationery Office, 1934. British History Online.

"The Deposition of Capt Thomas Dobbins late Comander of their Majties Ship Nonsuch now Commander of their Majties Gally called the Province Gally [. . .]." In "Affadavits of various sailors and officials relating to the complaints of Jahleel Brenton and Richard Short against Sir William Phips." Legal Document, TNA, Kew, CO 5/858. Adam Matthew, Colonial America Database.

Dinwiddie, Robert. "Report to the Board of Trade, April 1740." Qtd. in Kenneth Morgan, "Robert Dinwiddie's Reports on the British American Colonies." *William and Mary Quarterly* 3rd ser., 65, no. 2 (April. 2008): 305–46. JSTOR.

Dudley, Governor Joseph, to the Council of Trade and Plantations. July 13, 1704. *CSP*, vol. 22, 1704–1705, edited by Cecil Headlam, 211–23. London: His Majesty's Stationery Office, 1916. British History Online.

———. July 31, 1715. In *CSP*, vol. 28, 1714–15, edited by Cecil Headlam, 235–53. London: His Majesty's Stationery Office, 1928. British History Online.

"Establishment of a company of highland foot etc." Military Document, TNA, Kew, CO 5/654. Adam Matthew, Colonial America Database.

"Exceptions to the Province Accot of John Phillips Esq Late treasurer." In *The Acts and Resolves, Public and Private, of the Province of the Massachusetts Bay*, vol. 7. Boston: Wright & Potter Printing, 1892.

"Expedition Against the West Indies." Circulars. Sent to Connecticut, Maryland, Massachusetts, New Hampshire, New Jersey, New York, North Carolina, Rhode Island, and [Virginia?], April 2, 1740. In *Royal Instructions to British Colonial Governors, 1670–1776*, vol. 2, edited by Leonard Labaree. New York: D. Appleton, 1935.

"Extract from a Letter From Antigua." Qtd. in Vere Langford Oliver, *The History of the Island of Antigua, One of the Leeward Caribees in the West Indies [. . .]*, vol. 1. London: Mitchell and Hughes, 1891.

"Extract of a Letter from Mr. John Smith, on board the Success Frigate, Capt. Wm Thomson [. . .]. 14 July 1742." In *The Gentleman's Magazine and Historical Chronicle*, vol. 12, *For the Year M.DCCXLII*. London: Edward Cave, Junior, 1742.

"Extract of a [Letter] to Mr. Elisha Cook." In "Extracts and abstracts of letters from Boston." Correspondence, TNA, Kew, CO 5/855. Adam Matthew, Colonial America Database.

"Extract of Letter from Capt. Vernon to Mr. Burchett, 7 Nov. 1720." In *CSP*, vol. 32, 1720–1721, edited by Cecil Headlam, 329–46. London: His Majesty's Stationery Office, 1933. British History Online.

"Extracts from the statement of the services and expenses of the Province of

Massachusetts, made by a Committee of the Council and House of Representatives, chosen for the purpose in October, 1764, and sent to the colony's agent in England, to furnish arguments why the colony should not be taxed, &c." In *Speeches of the Governors of Massachusetts, From 1765 to 1775* [. . .], edited by Alden Bradford. Boston: Russell and Gardner, 1818.

Fane, Francis, to the Board of Trade. November 7, 1744. In "Documents concerning colonial finances and war with France." Correspondence, Legislation, Order, Report, TNA, Kew, CO 5/884. Adam Matthew, Colonial America Database.

Fenwicke, Colonel, to Captain Hardy. July 30, 1742. In "Council minutes relating to Captain Charles Hardy of HMS Rye." Minutes, Correspondence, Transcript, TNA, Kew, CO 5/369. Adam Matthew, Colonial America Database.

"From Benjamin Franklin to James Alexander and Cadwallader Colden with Short Hints towards a Scheme for Uniting the Northern Colonies, 8 June 1754." Founders Online, National Archives.

Gascoigne, James. Qtd. in W. E. May, "The Surveying Commission of Alborough, 1728–1734." *American Neptune* 21, no. 4 (October 1961).

Governor and Company of Rhode Island to the Council of Trade and Plantations. September 14, 1706. In *CSP*, vol. 23, *1706–1708*, edited by Cecil Headlam, 213–30. London: His Majesty's Stationery Office, 1916. British History Online.

Governor and Council of South Carolina to the Council of Trade and Plantations. October 21, 1718. In *CSP*, vol. 30, *1717–1718*, edited by Cecil Headlam, 359–81. London, His Majesty's Stationery Office 1930. British History Online.

Governor the Duke of Portland to the Council of Trade and Plantations. February 8, 1724/5. In *CSP*, vol. 34, *1724–1725*, edited by Cecil Headlam and Arthur Percival Newton, 320–35. London: His Majesty's Stationery Office, 1936. British History Online.

———. June 1, 1726. In *CSP, vol. 35, 1726–1727*, edited by Cecil Headlam and Arthur Percival Newton, 76–94. London: His Majesty's Stationery Office, 1936. British History Online.

Governor Parke's Reply to the 22 Articles of Complaint. June 26, 1709. In *CSP*, vol. 24, *1708–1709*, edited by Cecil Headlam, 370–408. London: His Majesty's Stationery Office, 1922. British History Online.

Habersham, Joseph, to William Henry Drayton. "Letter to William Henry Drayton Concerning Recruiting of Sailors for the Ship Prosper [. . .]." C. 1776. Robert W. Gibbes Collection of Revolutionary War Manuscripts (Series 213089, Box: 0004), South Carolina Department of Archives and History website.

Handasyd, Lieutenant Governor, to the Earl of Nottingham. December 10, 1702. Enclosing Minutes of Council of Jamaica, Dec. 4 and Dec. 7. In *CSP*, vol. 21, *1702–1703*, edited by Cecil Headlam, 13–44. London: His Majesty's Stationery Office, 1913. British History Online.

Hill, Lt. Thomas. "Remarks on board His Majesty's schooner, the St. John, in Newport harbor, Rhode Island." In *Records of the Colony of Rhode Island and Providence Plantations in New England*, vol. 6, edited by John Russell Bartlett. Providence: Knowles, Anthony, State Printers, 1861.

"His Honour the Lieutenant Governor's Speech at the Opening of a Sessions of the Council & Assembly of Jamaica on Tuesday the 27th day of September 1757." TNA, CO 137/30.

"The humble Address and Petition of the Governor and Councille and the Representatives of the Colony of the Massachusetts Bay; convened in General Court at Boston." In *The Andros Tracts: Being a Collection of Pamphlets and Official Papers*, vol. 2, edited by William Henry Whitmore. Boston: Prince Society, 1874.

"The Humble Peticon and Request of John Cooke." C. summer 1689. Massachusetts State Archives Collection, Colonial Period, 1622–1788, vol. 107, Revolution, 1689–1700, 79. FamilySearch website.

"The Humble Petition and Representation of the Council and Assembly of Your Majesty's Province of South Carolina Upon the Present State of the Said Province." July 26, 1740. In *Journal of the Commons House of Assembly of South Carolina, September 12, 1739–March 26, 1741*, edited by James H. Easterby. Columbia: University of South Carolina Press, 1952.

Hunter, Robert. Letter to the Council of Trade and Plantations and its inserts in America and West Indies: March 1733. In *CSP*, vol. 40, *1733*, edited by Cecil Headlam and Arthur Percival Newton, 16–31. London: His Majesty's Stationery Office, 1939. British History Online.

"An Impartial Narrative of ye Late Invasion of So Carolina by ye French + Spaniards, in the Month August 1706." In *Records in the British Public Record Office Relating to South Carolina 1701–1710*, edited by A. S. Salley. Columbia: Historical Commission of the State of South Carolina, 1947.

"Instructions for Governor Oglethorpe." October 9, 1739. Order, TNA, Kew, CO 5/654. Adam Matthew, Colonial America Database.

"Instructions to Coll. Vetch to be observed in his Negociations with the Gov'r of the several colonies dated March 1st. 1708/9." Order, TNA, Kew, CO 5/9. Adam Matthew, Colonial America Database.

Johnson, Governor Robert, to Colonel William Rhett. September 4, 1718. In "South Carolina Probate Records, Bound Volumes, 1671–1977," images. Charleston Wills, 1716–21, vol. 57, image 96 of 152. South Carolina Department of Archives and History, Columbia. FamilySearch website.

Johnson, Governor Robert, to Council of Trade and Plantations. January 12, 1720. In *CSP*, vol. 31, *1719–1720*, edited by Cecil Headlam, 293–311. London: His Majesty's Stationery Office, 1933. British History Online.

Laurens, Henry, to James Crokatt. December 28, 1747. In *The Papers of Henry Laurens*, vol. 1, *Sept. 11, 1746–Oct. 31, 1755*, edited by Philip M. Hamer, George C. Rogers, Maude E. Lyles. Columbia: University of South Carolina Press, 1968.

Lawes, Governor Nathaniel, to the Council of Trade and Plantations. January 31, 1719. In *CSP*, vol. 31, *1719–1720*, edited by Cecil Headlam, 1–21. London: His Majesty's Stationery Office, 1933. British History Online.

———. March 24, 1720. In *CSP*, vol. 31, *1719–1720*, edited by Cecil Headlam, 45–66. London: His Majesty's Stationery Office, 1933. British History Online.

———. November 13, 1720. *CSP*, vol. 32, *1720–1*, edited by Cecil Headlam, 187–95. London: His Majesty's Stationery Office, 1933. British History Online.

———. April 20, 1721. *CSP*, vol. 31, *1720–1721*, edited by Cecil Headlam, 281–97. London: His Majesty's Stationery Office, 1933. British History Online.

Le Jau, Francis, to the Society of the Propagation of the Gospel. Qtd. in Edgar Legare Pennington, "The South Carolina Indian War of 1715, as Seen by the Clergymen." *South Carolina Historical and Genealogical Magazine* 32, no. 4 (October 1931): 251–69. JSTOR.

"Letter concerning a sloop of war." May 8, 1691. Correspondence, TNA, Kew, CO 5/856. Adam Matthew, Colonial America Database.

"Letter from the Board of Trade to the Duke of Newcastle inclosing an extract from Mr. Bull, President of the Council in South Carolina, stating the danger that province is in from the row galleys which the Spaniards have at St. Augustine." Correspondence, TNA, Kew, CO 5/384. Adam Matthew, Colonial America Database.

"Letter from the Duke of Portland [. . .]." February 8, 1724/5. TNA, Kew, CO 137/16.

"Lieutenant Governor Bull to Duke of Newcastle and the affidavit of Thomas Lloyd and Robert Ford, Mariners, relating to their being taken by a Spanish sloop at St. Augustine, inclos'd." October 1741. Correspondence, TNA, Kew, CO 5/388. Adam Matthew, Colonial America Database.

"A list of the military strength of Carolina and Georgia." List, TNA, Kew, CO 5/655. Adam Matthew, Colonial America Database.

Lords of Trade and Plantations to King William III. June 12, 1690. In "New England patents and grants, 1690." Submission, Order, TNA, Kew, CO 5/905. Adam Matthew, Colonial America Database.

Lords Proprietors of Carolina to the Council of Trade and Plantations (No. 517). In *CSP*, vol. 28, edited by Cecil Headlam, 215–35. London: His Majesty's Stationery Office, 1928. British History Online.

Mead, Captain Samuel, to Josiah Burchett. December 27, 1715. TNA, Kew, ADM 1/2095.

"Memorial of the Lt. Governor and Council of the Massachusetts Bay to the King." July 8, 1726. In *CSP*, vol. 35, *1726–1727*, edited by Cecil Headlam and Arthur Percival Newton, 96–115. London: His Majesty's Stationery Office, 1936. British History Online.

Minot, John, to William Dummer. July 16, 1724. In *Letters of Colonel Thomas Westbrook and others relative to Indian affairs in Maine 1722–1726*, edited by William Blake Trask. Boston: G. E. Littlefield, 1901.

Montagu, Admiral, to Governor Wanton. April 8, 1772. In *Records of the Colony of Rhode Island and Providence Plantations in New England*, vol. 7, edited by John Russell Bartlett. Providence: Knowles, Anthony, State Printers, 1861.

"Mr. Pownall's Consideration Towards a General Plan of Measures for the Colonies." In Matthew S. Quay, *Pennsylvania Archives, Second Series, vol. 6*. Harrisburg: Lane S. Hart, 1877.

"News from New England." February 2, 1690/1. In "Extracts and news from New [England]." Correspondence, TNA, Kew, CO 5/856. Adam Matthew, Colonial America Database.

Oglethorpe, General, to Harman Verelst. December 7, 1741. In *Trustees for Establishing the Colony of Georgia in America, "Letters from Georgia, v. 14206, 1741 June–1742 December."* Digital Library of Georgia.

"Orders for John Cooke [. . .]." Massachusetts State Archives, vol. 127, 266. FamilySearch website.

"Petition of Governor Philipps to the King." In *CSP*, vol. 34, *1724–1725*, edited by Cecil Headlam and Arthur Percival Newton, 86–105. London: His Majesty's Stationery Office, 1936. British History Online.

Philipps, Governor Richard, to the Council of Trade and Plantations. August [6?], 1720. In *CSP*, vol. 32, *1720–1721*, edited by Cecil Headlam, 77–97. London: His Majesty's Stationery Office, 1933. British History Online.

———. September 27, 1720 (and attachments). In *CSP*, vol. 32, *1720–1721*, edited by Cecil Headlam, 144–65. London: His Majesty's Stationery Office, 1933. British History Online.

———. August 16, 1721. in *CSP*, vol. 32, *1720–1721*, edited by Cecil Headlam, 388–402. London: His Majesty's Stationery Office, 1933. British History Online.

Pownall, Governor Thomas, to Pitt. August 16, 1757. In *Correspondence of William Pitt When Secretary of State, vol. 1*, edited by Gertrude Selwyn Kimball. New York: Macmillan, 1906.

Proceedings and Debates of the British Parliaments Respecting North America. Vol. 5, *1739–1754*, edited by Leo Francis Stock. Washington, DC: Carnegie Institution of Washington, 1941.

Quary, Robert, to the Council of Trade and Plantations. December 7, 1702. In *CSP*, vol. 21, *1702–1703*, edited by Cecil Headlam, 13–44. London: His Majesty's Stationery Office, 1913. British History Online.

"Reasons and Motives for the Albany Plan of Union, [July 1754]." Founders Online, National Archives.

"Representation to the King on the Proceedings of the Congress at Albany." In *Documents Relative to the Colonial History of the State of New-York*, vol. 6, edited by E. B. O'Callaghan. Albany: Weed, Parsons, 1855.

Rogers, Governor Woodes, to the Council of Trade and Plantations. October 31, 1718. In *CSP*, vol. 30, *1717–1718*, edited by Cecil Headlam, 359–81. London: His Majesty's Stationery Office, 1930. British History Online.

Shirley, Governor William, to the Duke of Newcastle. March 27, 1745. In *Correspondence of William Shirley, Governor of Massachusetts and Military Commander in America, 1731–1760*, edited by Charles Henry Lincoln. New York: Macmillan, 1912.

Shirley, William, to the Lords of Trade. December 1, 1747. In *Correspondence of William Shirley, Governor of Massachusetts and Military Commander in America, 1731–1760*, edited by Charles Henry Lincoln. New York: Macmillan, 1912.

"Sir Edmund Andros' account of the force raised in the year 1688 for the defence of New England against the Indians." Report, Military Document, TNA, Kew, CO 5/855. Adam Matthew, Colonial America Database.

Sparhawk, Nathaniel, to William Pepperell. January 14, 1745/6. In *Collections of the*

Massachusetts Historical Society, Sixth Series, vol. 10. Boston: Massachusetts Historical Society, 1899.

Stoughton, Lt. Governor William, to Council of Trade and Plantations. September 30, 1697. In *CSP*, vol. 15, *15 May, 1696–31 October 1697*, edited by J. W. Fortescue, 624–25. London: Mackie, 1904. British History Online.

Vetch, Col. Samuel, to [?the Earl of Sunderland]. August 2 and 12, 1709. In *CSP*, vol. 24, *1708–1709*, edited by Cecil Headlam, 437–57. British History Online.

Wanton, Governor, to Admiral Montagu. May 8, 1772. In *Records of the Colony of Rhode Island and Providence Plantations in New England*, vol. 7, edited by John Russell Bartlett. Providence: Knowles, Anthony, State Printers, 1861.

Warren, Peter, to George Anson. April 2, 1745. In *The Royal Navy and North America: The Warren Papers, 1736–1752*, edited by Julian Gwyn. Navy Records Society, 1973.

Warren, Peter, to George Clinton. July 6, 1744. In *The Royal Navy and North America: The Warren Papers, 1736–1752*, edited by Julian Gwyn. Navy Records Society, 1973.

Worsley, Governor Henry, to the Council of Trade and Plantations. March 26, 1726. In *CSP*, vol. 33, *1722–1723*, edited by Cecil Headlam, 221–38. London: His Majesty's Stationery Office, 1934. British History Online.

Legislative Minutes/Records

"27 August." In "Volume 244: July 1–December 30, 1723," *Calendar of Treasury Papers*, vol. 6, *1720–1728*, edited by Joseph Redington, 218–37. London, 1889. British History Online.

The Acts and Resolves, Public and Private, of the Province of the Massachusetts Bay. Vol. 3. Boston: Albert J. Wright, 1878.

The Acts and Resolves, Public and Private, of the Province of the Massachusetts Bay. Vol. 8, *Being Volume III of the Appendix, Containing Resolves, Etc. 1703–1707*, Boston: Wright & Potter Printing, 1895.

Calendar of Historical Manuscripts, in the Office of the Secretary of State. Part 2, edited by E. B. O'Callaghan. Albany: Weed, Parsons, Printers, 1866.

"Copy of Act of Jamaica for Fitting out Sloops for Guarding the Coasts etc." In *CSP*, vol. 32, *1720–1721*, edited by Cecil Headlam, 329–46. London: His Majesty's Stationery Office, 1933. British History Online.

Council and Council in Assembly Minutes, Jamaica. September 29, 1712. TNA, Kew, CO 140/11.

———. September 7, 1716. TNA, Kew, CO 140/13.

———. January 8, 1718/19. TNA, CO 140/16.

Council Minutes, Dominion of New England. May 25, 1687. In "Proceedings of the Council of the Dominion of New England from 4th May to 28th July 1687." Minutes, TNA, Kew, CO 5/785. Adam Matthew, Colonial America Database.

———. July 28, 1687. In "Proceedings of the Council of the Dominion of New England from 4th May to 28th July 1687." Minutes, TNA, Kew, CO 5/785. Adam Matthew, Colonial America Database.

Council Minutes, Jamaica. April 7–9, 1729. TNA, CO 140/21.

Council Minutes, Massachusetts. May 2, 1701. In "Minutes of the council of the

Massachusetts Bay from 9th January to 13th May 1701." Minutes, TNA, Kew, CO 5/788. Adam Matthew, Colonial America Database.

———. September 4, 1701. "Minutes of the council of the Massachusetts Bay from 30th May to 17th September 1701." Minutes, TNA, Kew, CO 5/788. Adam Matthew, Colonial America Database.

———. October 9, 1704. "Minutes of the Massachusetts council, Jun 1704–Mar 1705." Minutes, TNA, Kew, CO 5/789. Adam Matthew, Colonial America Database.

———. July 19–August 10, 1710. In "Minutes of council of the Massachusetts Bay, Jul–Nov 1710." Minutes, TNA, Kew, CO 5/791. Adam Matthew, Colonial America Database.

———. June 7, 1715. In "Massachusetts: Minutes of Council 21 Mar–11 Oct 1715." Minutes, TNA, Kew, CO 5/792. Adam Matthew, Colonial America Database.

———. June 6, 1722. In "Minutes of Council of the Province of the Massachusetts Bay 2 Mar–20 Aug 1722." Minutes, TNA, Kew, CO 5/794. Adam Matthew, Colonial America Database.

———. June 25–26, 1722. In "Minutes of Council of the Province of the Massachusetts Bay 2 Mar–20 Aug 1722." Minutes, TNA, Kew, CO 5/794. Adam Matthew, Colonial America Database.

———. June 29, 1722. In "Minutes of Council of the Province of the Massachusetts Bay 2 Mar–20 Aug 1722." Minutes, TNA, Kew, CO 5/794. Adam Matthew, Colonial America Digital Database.

———. July 25, 1722. In "Minutes of Council of the Province of the Massachusetts Bay 2 Mar–20 Aug 1722." Minutes, TNA, Kew, CO 5/794. Adam Matthew, Colonial America Digital Database.

———. June 28–29, 1726. In "Minutes of the Council of Massachusetts Bay." Minutes, TNA, Kew, CO 5/797. Adam Matthew, Colonial America Database.

Council Minutes, Pennsylvania. June 3 and October 16, 1741. In *Minutes of the Provincial Council of Pennsylvania, from the Organization to the Termination of the Proprietary Government*, vol. 4. Harrisburg: Theo. Fenn, 1851.

———. June 14, 1748. In *Minutes of the Provincial Council of Pennsylvania, from the Organization to the Termination of the Proprietary Government*, vol. 5. Harrisburg, Theo. Fenn, 1851.

Council Minutes, South Carolina. July 19, 1742. In "Council minutes relating to Captain Charles Hardy of HMS Rye." Minutes, Correspondence, Transcript, TNA, Kew, CO 5/369. Adam Matthew, Colonial America Database.

Journal of Assembly of Barbados. August 25, 1702. In *CSP*, vol. 20, 1702, edited by Cecil Headlam, 548–66. London: His Majesty's Stationery Office, 1912. British History Online.

———. September 19, 1702. In *CSP*, vol. 20, 1702, edited by Cecil Headlam, 592–99. London: His Majesty's Stationery Office, 1912. British History Online.

Journals of the House of Commons. Vol. 25. London: House of Commons, 1803.

Legislature Minutes, Georgia. February 10, 1778. In *The Revolutionary Records of the State of Georgia*, vol. 2, edited by Allen D. Candler. Atlanta: Franklin-Turner, 1908.

Legislature Minutes, Massachusetts. May 14, 1645. In *Records of the Governor and*

Company of the Massachusetts Bay in New England, vol. 3, *1644–1657*, edited by Nathaniel B. Shurtleff. Boston: William White, 1854.

———. June 4, 1717. In *Journals of the House of Representatives of Massachusetts, 1715–1717*. Boston: Massachusetts Historical Society, 1919.

———. June 27–28, 1722. In *Journals of the House of Representatives of Massachusetts*, vol. 4, *1722–1723*. Boston: Massachusetts Historical Society, 1923.

———. August 27, 1726. In *Journals of the House of Representatives of Massachusetts, 1726–1727*. Boston: Massachusetts Historical Society, 1926.

Legislature Minutes, Pennsylvania. November 24, 1758. In *Pennsylvania Archives, Eighth Series*, vol. 6, *October 14, 1756–January 3, 1764*, edited by Charles F. Hoban. Philadelphia: Pennsylvania State Library, 1935.

Legislature Minutes, South Carolina (Upper House of Assembly and the Commons House of Assembly). *Journal of the Commons House of Assembly.*

———. September 1, 1727. Commons House Minutes, TNA, Kew, CO 5/429, 225–227. Adam Matthew, Colonial America Digital Database.

———. April 4, 1728. Commons House Minutes, TNA, Kew, CO 5/430. Adam Matthew, Colonial America Digital Database.

———. July 10–13, 1728. Upper House Minutes. In "Proceedings in the Upper House of assembly." Minutes, TNA, Kew, CO 5/429. Adam Matthew, Colonial America Database.

———. November 8, 1739. Commons House Minutes. In *Journal of the Commons House of Assembly, September 12, 1739–March 26, 1741*, edited by J. H. Easterby. Columbia: Historical Commission of South Carolina, 1952.

———. April 2, 1740. Commons House Minutes. In *Journal of the Commons House of Assembly, September 12, 1739–March 26, 1741*, edited by J. H. Easterby.

———. July 18, 1740. Commons House Minutes. In *Journal of the Commons House of Assembly, September 12, 1739–March 26, 1741*, edited by J. H. Easterby.

———. November 19–22 and December 1, 1742. Commons House Minutes. In *Journal of the Commons House of Assembly, September 14, 1742–January 27, 1744*, edited by J. H. Easterby. Columbia: South Carolina Archives Department, 1954.

"Massachusetts Documents, 1689–1692." Edited by Robert Moody, 252–55. Colonial Society of Massachusetts website.

Minutes of the Council [in Assembly] of Barbados. March 26–27, 1667. In *CSP*, vol. 5, *1661–1668*, edited by W. Noel Sainsbury, 451–59. London: Her Majesty's Stationery Office, 1880. British History Online.

———. August 25–26 and September 8, 1702. In *CSP*, vol. 20, *1702*, edited by Cecil Headlam, 581–88. London: His Majesty's Stationery Office, 1912. British History Online.

———. September 15, 1702. In *CSP*, vol. 20, *1702*, edited by Cecil Headlam, 588–92. London: His Majesty's Stationery Office, 1912. British History Online.

———. September 18, 1702. In *CSP*, vol. 20, *1702*, edited by Cecil Headlam, 592–99. London: His Majesty's Stationery Office, 1912. British History Online.

"Order on Committees Report About Ye Defence of the Coast Agst Pyrates." In *The Acts and Resolves, Public and Private, of the Province of the Massachusetts Bay*. Vol. 10, *Resolves, Etc. 1720–1726*. Boston: Wright & Potter Printing State Printers, 1902.

The Public Records of the Colony of Connecticut, From May, 1757, to March, 1762, Inclusive. Edited by Charles J. Hoadly. Hartford: Case, Lockwood, & Barnard, 1880.

Records of the Colony of New Plymouth in New England. Vol. 2, *Acts of the Commissioners of the United Colonies of New England, 1653–1679*, edited by David Pulsifer. Boston: William White, 1859.

Records of the Colony of Rhode Island and Providence Plantations. Vol. 6, *1757 to 1769*, edited by John Russell Bartlett. Providence: Knowles, Anthony, 1861.

"Vote for Encouraging the Prosecution of Ye Pyrates." In *The Acts and Resolves, Public and Private, of the Province of the Massachusetts Bay.* Vol. 10, *Resolves, Etc. 1720–1726.* Boston: Wright & Potter Printing, State Printers, 1902.

"Vote for Fitting out Two Shallops Against the Indians." August 18, 1722. In *The Acts and Resolves, Public and Private, of the Province of the Massachusetts Bay.* Vol. 10, *Resolves, Etc., 1720–1726.* Boston: Wright & Potter Printing, 1902.

"Vote relating to the Prov: Privateer, June 11 1740." Massachusetts State Archives Collection, Colonial Period, vol. 62, 596. FamilySearch website.

Newspapers

American Weekly Mercury (Philadelphia)
Boston Gazette
Boston News-Letter
London Gazette
New-England Weekly Journal (Boston)
New Hampshire Gazette (Portsmouth)
Newport (RI) Mercury
New-York Mercury
Pennsylvania Gazette (Philadelphia)
Publick Occurrences (Boston)
South Carolina Gazette (Charles Town)

Royal Navy Ship Logs

Logbook of HMS *Flamborough*. TNA, ADM 51/357
Logbook of HMS *Flamborough*. TNA, ADM 51/358
Logbook of HMS *Sheerness*. TNA, ADM 51/898

Admiralty Court Records

"The Appellants Case." In *Prize Appeals, 1736–1751*, vol. 1, fol. 75–76, edited by Sir George Lee. New York Public Library Digital Collections.

"The Case of His Majesty's Ships Chester and Sunderland, the actual and sole Captors of the Prize." July 5, 1750. In *Prize Appeals, 1736–1751*, vol. 1, fol. 65, edited by Sir George Lee. New York Public Library Digital Collections.

"Notes of Arguments Advanced by Dr. Pinfold and Mr. Yorke." May 17, 1749. In *Prize Appeals, 1751–1758*, vol. 2, fol. 256, edited by Sir George Lee. New York Public Library Digital Collections.

"Notre Dame de Deliverance . . . The Case of the Three Respondents." May 3, 1750. In

Prize Appeals, 1736–1751, vol. 1, fol. 65, edited by Sir George Lee. New York Public Library Digital Collections.

Records of the South Carolina Court of Admiralty, 1716–1732. National Archives, Washington, DC, N.D., microfilm, pp. 306–90. Accessed at Charleston County Library, South Carolina History Room.

"Richardson & others V. Ship Two Friends & Cargo, Decree." In Suffolk County (MA) Court Files, 1629–1797, v. 384, case 61447. FamilySearch website.

Books, Pamphlets, Broadsides

Bull, Governor William, to Col. Vanderdussen. July 9, 1740. In *Report of the Committee Appointed by the General Assembly [. . .]*. Vol. 4, Charleston: Walker, Evans & Cogswell, 1887.

Church, Benjamin. *The History of King Philip's war; also of expeditions against the French and Indians in the eastern parts of New-England, in the years 1689, 1690, 1692, 1696 and 1704 [. . .]*. Repr. ed.; Boston: Howe & Norton, 1825.

Colonial Records of the State of Georgia. Vol. 4, *Stephens' Journal, 1737–40*. Edited by Allen D. Candler. Atlanta: Franklin Printing and Publishing, 1906.

Faulkner, George. *The Present State of the Revenues and Forces by Sea and Land Of France and Spain, Compared with those of Great Britain [. . .]*. Dublin: George Faulkner, 1740.

Franklin, Benjamin. "Plain Truth, 17 November 1747." Founders Online, National Archives.

The Gentleman's Magazine And Historical Chronicle. Vol. 10, *For the Year M.DCCXL* London: Edw. Cave, 1740.

———. Vol. 20, *For the Year MDCCL*. London: Edward Cave, 1750.

Hamilton, Lord Archibald. *An Answer to an Anonymous Libel, Entitled, Articles Exhibited Against Lord Archibald Hamilton, Late Governour of Jamaica: With Sundry Depositions and Proofs Relating to the Same*. London, 1718.

Hopkins, Stephen, *The Rights of the Colonies Examined*. Providence: William Goddard, 1765. Evans Early Imprint Collection.

An Impartial Account of the Late Expedition Against St. Augustine Under General Oglethorpe 1742. Repr. ed. Gainesville: University Presses of Florida, 1978.

Johnson, Charles. *A General History of the Pirates, from Their First Rise and Settlement in the Island of Providence, to the Present Time*. London: T. Warner, 1724.

Johnson, Samuel. *A Dictionary of the English Language: A Digital Edition of the 1755 Classic*. 1755; repr., Johnson Dictionary Online, 2012.

Journal of the Commissioners for Trade and Plantations, From January 1741-2 to December 1749. London: His Majesty's Stationery Office, 1931.

Mather, Increase. "The Present State of the New English Affairs." September 3, 1689. In *The Andros Tracts: Being a collection of pamphlets and official papers issued during the period between the overthrow of the Andros government and the establishment of the second charter of Massachusetts*, vol. 3, edited by William Henry Whitmore. Boston: Prince Society, 1868–74.

———. "A Vindication of New England." In *The Andros Tracts: Being a Collection of*

Pamphlets and Official Papers [. . .], vol. 6, edited by William Henry Whitmore. Boston: Prince Society, 1869.

Otis, James. *A Vindication of the Conduct of the House of Representatives of the Province of the Massachusetts-Bay* [. . .]. Boston: Edes & Gill, 1762.

Penhallow, Samuel. *The History of the Wars of New-England with the Eastern Indians.* 1726; repr., Boston: Oscar H. Harpel, Chestnut Street, 1859.

Phillips, Edward. *The New World of Words: Or Universal English Dictionary.* London: King's Arms, 1720.

Report of the Committee Appointed by the General Assembly of South Carolina in 1740 on the St. Augustine Expedition Under General Oglethorpe, Collections of the Historical Society, of South Carolina. Vol. 4, Charleston: Walker, Evans & Cogswell, 1887.

The Scots Magazine . . . For the Year MCCCXLII. Vol. 4. Edinburgh: Sands, Brymer, Murray and Cochran.

The State of the Island of Jamaica. Chiefly in Relation to its Commerce and the Conduct of the Spaniards in the West-Indies. London: H. Whitridge, 1726.

Vetch, Samuel. "Canada Survey'd [. . .]." July 27, 1708. In *CSP*, vol. 24, 1708–1709, edited by Cecil Headlam, 40–56. London: His Majesty's Stationery Office, 1922. British History Online.

Winthrop, John. *Winthrop's Journal, History of New England, 1630–1649.* Vol. 1. Edited by James Kendall Hosmer. New York: Charles Scribner's Sons, 1908.

Yonge, Francis. *A Narrative of the Proceedings of the People of South-Carolina, in the Year 1719: And of the True Causes and Motives that Induced Them to Renounce Their Obedience to the Lords Proprietors, as Their Governors, and to Put Themselves Under the Immediate Government of the Crown.* Vol. 1. London, 1726.

Secondary Sources
Academic Books and Journal Articles

Alsop, James D. "Samuel Vetch's 'Canada Survey'd': The Formation of a Colonial Strategy, 1706–1710." *Acadiensis* 12, no. 1 (1982): 39–58. JSTOR.

Anderson, Fred. *Crucible of War: The Seven Years' War and the Fate of Empire in British North America, 1754–1766.* New York: Alfred A. Knopf, 2000.

Andrews, Charles McLean, ed. "Introduction." In *Original Narratives of Early American History: Narratives of the Insurrections 1675–1690.* New York: Charles Scribner's Sons, 1915.

Andrews, Kenneth R. *Ships, Money, and Politics: Seafaring and Naval Enterprise in the Reign of Charles I.* Cambridge: Cambridge University Press, 1991.

Armitage, David. *The Ideological Origins of the British Empire.* Cambridge: Cambridge University Press, 2000. ProQuest.

Arnade, Charles W. *The Siege of St. Augustine in 1702.* Gainesville: University of Florida Press, 1959.

Arnold, Samuel Greene. *History of the State of Rhode Island and Providence Plantations*, vol. 2, 1700–1790. New York: D. Appleton, 1860.

Austin, John Osborne, *The Genealogical Dictionary of Rhode Island.* Albany: Joel Munsell's Sons, 1887.

Bacon, Edwin Monroe. *The Connecticut River and the Valley of the Connecticut Three Hundred and Fifty Miles from Mountain to Sea; Historical and Descriptive*. New York: G. P. Putnam's Sons, 1906.

Bahar, Matthew. "People of the Dawn, People of the Door: Indian Pirates and the Violent Theft of an Atlantic World." *Journal of American History* 101, no. 2 (September 2014): 401–26.

———. *Storm of the Sea: Indians and Empires in the Atlantic's Age of Sail*. Oxford: Oxford University Press, 2019. Kindle eBook edition.

Baine, Rodney E., "General James Oglethorpe and the Expedition against St. Augustine." *Georgia Historical Quarterly* 84, no. 2 (Summer 2000): 197–229. JSTOR.

Baker, William Avery. "Vessel Types of Colonial Massachusetts." In *Collections of the Colonial Society of Massachusetts*, vol. 52, *Sea Faring in Colonial Massachusetts* (March 1980), 18–20. Colonial Society of Massachusetts website.

Baugh, Daniel. *British Naval Administration in the Age of Walpole*. Princeton, NJ: Princeton University Press, 1965.

———. *The Global Seven Years War, 1754–1763*. Abingdon: Routledge, 2011.

———. "Great Britain's 'Blue-Water' Policy, 1689–1815," *International History Review* 10, no. 1 (1988): 33–58. JSTOR.

———, ed. *Naval Administration, 1715–1750*. London: Navy Records Society, 1977.

Bazan, Daniel R. *For Want of Sloops, Water Casks, and Rum: The Difficulty of Logistics in the Canadian Theater of the Seven Years War*. Unpublished master's thesis, Liberty University, 2013. Liberty University Digital Commons.

Beaumont, Andrew D. M. *Colonial America and the Earl of Halifax*. Oxford: Oxford University Press, 2015.

Black, Jeremy, *America or Europe?: British Foreign Policy, 1739–63*. London: Taylor & Francis Group, 1997.

Bourne, Ruth. *Queen Anne's Navy in the West Indies*. New Haven: University of New Haven, 1939.

Boyer, Paul, and Stephen Nissenbaum, *Salem Possessed*. Cambridge: Harvard University Press, 1974. JSTOR.

Braddick, Michael J. *State Formation in Early Modern England, c. 1550–1700*. Cambridge: Cambridge University Press, 2000. ProQuest.

Braddock, J. G., "The Plight of a Georgia Loyalist: William Lyford, Jr." *Georgia Historical Quarterly* 91, no. 3 (2007): 247–65.

Bradley, Peter T. *British Maritime Enterprise in the New World from the Late Fifteenth to the Mid-Eighteenth Century*. Lampeter: Edwin Mellen Press, 1999.

Braudel, Fernand. *Civilization and Capitalism, 15th–18th Century*. Vol. 2, *The Wheels of Commerce*, translated by Sian Reynolds. Berkeley: University of California Press.

Breen, Louise A. *Transgressing the Bounds: Subversive Enterprises among the Puritan Elite in Massachusetts, 1630–1692*. Cary: Oxford University Press, 2001. ProQuest.

Brewer, John, *The Sinews of Power: War, Money and the English State, 1688–1783*. London: Unwin Hyman, 1989.

Brunsman, Denver. *The Evil Necessity: British Naval Impressment in the Eighteenth-Century Atlantic World*. Charlottesville: University of Virginia Press, 2013. ProQuest.

Buchet, Christian. "The Royal Navy and the Caribbean, 1689–1763." *Mariner's Mirror* 80, no. 1 (1994): 30–44. Taylor and Francis Digital Access.

Buchet, Christian, Anita Higgie, and Michael Duffy. *The British Navy, Economy and Society in the Seven Years War*. Woodbridge, Suffolk; Rochester, NY: Boydell & Brewer, 2013. JSTOR.

Buker, George E., and Richard Apley Martin. "Governor Tonyn's Brown-Water Navy: East Florida during the American Revolution, 1775–1778." *Florida Historical Quarterly* 58, no. 1 (1979): 58–71. JSTOR.

Burgess, Douglas R., Jr. *The Politics of Piracy: Crime and Civil Disobedience in Colonial America*. Lebanon: University Press of New England, 2014. ProQuest.

Butler, Dr. Nic. "Anson and the Privateer Emergency of 1727." Unpublished MSS/collection of research from SC Archives graciously shared with author.

Carr, J. Revell. *Seeds of Discontent: The Deep Roots of the American Revolution, 1650–1750* New York: Bloomsbury, 2008. Kindle eBook Edition.

Cate, Margaret Davis. "Fort Frederica and the Battle of Bloody Marsh." *Georgia Historical Quarterly* 27, no. 2 (June 1943): 111–74. JSTOR.

Chapin, Howard. "New England Vessels in the Expedition against Louisbourg, 1745." *New England Historical and Genealogical Register* 76 (January 1922).

———. *Privateering in King George's War, 1739–1748*. Providence: E. A. Johnson, 1928.

———. *Privateer Ships and Sailors: The First Century of American Colonial Privateering, 1625–1725*. 1926; repr., Martino Fine Books, 2017.

———. *The Tartar: The Armed Sloop of the Colony of Rhode Island in King George's War*. Providence: Society of Colonial Wars, 1922.

Coakley, John. "'The Piracies of Some Little Privateers': Language, Law and Maritime Violence in the Seventeenth-Century Caribbean." *Britain and the World* 13, no. 1 (2020): 6–26. EBSCOhost.

Coker, P. C. *Charleston's Maritime Heritage, 1670–1865: An Illustrated History*, Coker Craft, 1987.

Coleman, Emma Lewis. *New England Captives Carried to Canada between 1677 and 1760 during the French and Indian Wars*. Vol. I. Heritage Books, 2008.

Crane, Verner W. *The Southern Frontier, 1670–1732*. Durham: Duke University Press, 1928.

Crittenden, Charles Christopher, "The Surrender of the Charter of Carolina." *North Carolina Historical Review* 1, no. 4 (October 1924): 383–402. JSTOR.

Cumberland, Barlow. *History of the Union Jack and Flags of the Empire*. W. Briggs, 1909.

David, Huw. *Trade, Politics, and Revolution: South Carolina and Britain's Atlantic Commerce, 1730–1790*. Columbia: University of South Carolina Press, 2018. JSTOR.

Davies, J. D. *Pepys's Navy: Ships, Men & Warfare, 1649–1689*. Barnsley: Seaforth Publishing, 2008. Kindle eBook Edition.

Daughn, George C. *If By Sea: The Forging of the American Navy: From the American Revolution to the War of 1812*. New York: Basic Books, 2008. Kindle eBook Edition.

De Forest, Louis Effingham, ed. "Appendix I: The Fleet." In *Louisbourg Journals 1745*. New York: Society of Colonial Wars in New York, 1932.
Derderian, Michael "This Licentious Republic: Maritime Skirmishes in Narragansett Bay 1763–1769." *Journal of the American Revolution*, Electronic Journal, October 2, 2017.
Douglas, W. A. B. *Nova Scotia and the Royal Navy, 1713–1766*. Unpublished diss., Queen's University, 1973. Microfilm.
———. "The Sea Militia of Nova Scotia, 1749–1755: A Comment on Naval Policy." *Canadian Historian Review* 47, no. 1 (March 1966): 22–37.
Dow, George Francis Dow, and John Henry Edmonds. *Pirates of the New England Coast, 1630–1730*. Salem: Marine Research Society, 1923.
Dull, John R., *American Naval History, 1607–1865: Overcoming the Colonial Legacy*. Lincoln: University of Nebraska Press, 2012.
———. *The French Navy and the Seven Years War*. Lincoln: University of Nebraska Press, 2005.
Dunn, Richard S. "The Glorious Revolution and America." In *Origins of Empire: British Overseas Enterprise to the Close of the Seventeenth Century*, edited by Nicholas Canny. Oxford: Oxford University Press, 2001. ProQuest.
———. "The Trustees of Georgia and the House of Commons, 1732–1752," *William and Mary Quarterly* 11, no. 4 (October 1954): 551–65. JSTOR.
Earle, Peter. *The Pirate Wars*. Macmillan, 2003. Kindle eBook edition.
Eddison, Jill. *Medieval Pirates: Pirates, Raiders and Privateers, 1204–1453*. Stroud: History Press, 2013. Kindle eBook edition.
Edgar, Walter. *South Carolina: A History*. Columbia: University of South Carolina Press, 1998.
Edmonds, John Henry. *Captain Thomas Pound*. Cambridge: John Wilson and Son, 1918.
Egle, William H. *History of the Commonwealth of Pennsylvania, Civil, Political, and Military, from Its Earliest Settlement to the Present Time*. Philadelphia: E. M. Gardner, 1883.
Elizas, Barnett A. *The Jews of South Carolina: From the Earliest Times to the Present Day*. Philadelphia: J. B. Lippincott, 1905.
Ettinger, Amos Aschbach. *James Edward Oglethorpe, Imperial Idealist*. Oxford: Clarendon Press, 1936.
Finucane, Adrian. *The Temptations of Trade: Britain, Spain, and the Struggle for Empire*. Philadelphia: University of Pennsylvania Press, 2016. ProQuest.
Fox, E. T. "Jacobitism and the 'Golden Age' of Piracy, 1715–1725." *International Journal of Maritime History* 22, no. 2 (December 2010): 277–303.
Gallay, Allan. *The Indian Slave Trade: The Rise of the English Empire in the American South, 1670–1717*. New Haven: Yale University Press, 2002. ProQuest.
Gilje, Paul A. *Liberty on the Waterfront: American Maritime Culture in the Age of Revolution*. Philadelphia: University of Pennsylvania Press, 2007. ProQuest.
Glete, Jan. *Warfare at Sea, 1500–1650: Maritime Conflicts and the Transformation of Europe*. London: Taylor & Francis Group, 1999. ProQuest.
Godfrey, William C. *Pursuit of Profit and Preferment in Colonial North America: John Bradstreet's Quest*. Waterloo: Wilfrid Laurier University Press, 1982.

Goldberg, Dror. "The Massachusetts Paper Money of 1690." *Journal of Economic History* 69, no. 4 (2009): 1092–1106. JSTOR.

Gould, Eliga. *Among the Powers of the Earth: The American Revolution and the Making of a New World Empire.* Cambridge: Harvard University Press, 2012. ProQuest.

———. *The Persistence of Empire: British Political Culture in the Age of the American Revolution.* Chapel Hill: University of North Carolina Press, 2000. ProQuest.

Graham, G. S. "The Naval Defence of British North America, 1739–1763." *Transactions of the Royal Historical Society* 30 (1948): 95–110. JSTOR.

Grainger, John D. *The British Navy in the Caribbean.* Woodbridge: Boydell & Brewer, 2021. Kindle eBook edition.

Greene, Jack. *The Constitutional Origins of the American Revolution.* New Histories of American Law. Cambridge: Cambridge University Press, 2011. ProQuest.

———. *Peripheries and Center: Constitutional Development in the Extended Polities of the British Empire and the United States, 1607–1788.* New York: Norton, 1990.

Grenier, John. *The Far Reaches of Empire: War in Nova Scotia, 1710–1760.* Norman: University of Oklahoma, 2008.

Gwyn, Julian. *Frigates and Foremasts: The North American Squadron in Nova Scotia Waters 1745–1815.* Vancouver: University of British Columbia Press, 2003.

Hahn, Steven C. "The Atlantic Odyssey of Richard Tookerman: Gentleman of South Carolina, Pirate of Jamaica, and Litigant before the King's Bench." *Early American Studies* 15, no. 3 (Summer 2017): 539–90. JSTOR.

Hanna, Mark. *Pirate Nests and the Rise of the British Empire, 1570–1740.* Chapel Hill: University of North Carolina Press, 2015. ProQuest.

Harding, Richard. *The Emergence of Britain's Global Naval Supremacy: The War of 1739–1748.* Woodbridge: Boydell and Brewer, 2010. JSTOR.

———. *The Evolution of the Sailing Navy, 1509–1815.* New York: St. Martin's Press, 1995.

———. *Seapower and Naval Warfare, 1650–1830.* Routledge, 1999. ProQuest.

———. "The War in the West Indies." In *The Seven Years' War: Global Views*, edited by Mark Danley and Patrick Speelman. Leiden: BRILL, 2012. ProQuest.

Harris, Lynn. *Patroons and Periaguas: Enslaved Watermen and Watercraft of the Lowcountry* Columbia: University of South Carolina Press, 2014. EBSCOhost.

Hattendorf, John. *Talking About Naval History: A Collection of Essays.* Pittsburgh: US Government Printing Office, 2012.

Herson, Major James P. *A Joint Opportunity Gone Awry: The 1740 Siege of St. Augustine.* Fort Leavenworth: United States Army Command and General Staff College, 1997.

Higginbotham, Don. "The Early American Way of War: Reconnaissance and Appraisal." *William and Mary Quarterly* 44, no. 2 (April 1987): 230–73. JSTOR.

Hitchings, Sinclair. "Guarding the New England Coast: The Naval Career of Cyprian Southack." In *Publications of the Colonial Society of Massachusetts*, vol. 52, *Seafaring in Colonial Massachusetts, A Conference Held by the Colonial Society of Massachusetts November 21 and 22, 1975, Boston: The Colonial Society of Massachusetts, 1980.* Colonial Society of Massachusetts website.

Ivers, Larry E. *British Drums on the Southern Frontier: The Military Colonization of Georgia, 1733–1749.* Chapel Hill: University of North Carolina Press, 1974.

———. *This Torrent of Indians: War on the Southern Frontier, 1715–1728*. Columbia: University of South Carolina Press, 2016. Kindle eBook edition.

Jabbs, Theodore. *South Carolina Colonial Militia, 1663–1733*. PhD diss., University of North Carolina, 1973.

Jarvis, Michael. *In the Eye of All Trade: Bermuda, Bermudians, and the Maritime Atlantic World, 1680–1783*. Chapel Hill: Omohundro Institute of Early American History & Culture, 2010.

Jasanoff. Maya, *Liberty's Exiles: American Loyalists in the Revolutionary World*. New York: Vintage Books, 2011.

Jestice, Phyllis. "Naval Warfare." In *Fighting Techniques of the Medieval World, AD 500–AD 1500*, edited by Matthew Bennett, Jim Bradbury, Kelly DeVries, Iain Dickie, and Phyllis Jestice. New York: St. Martin's Press, 2005.

Johnson, A. J. B. *Endgame 1758: The Promise, the Glory, and the Despair of Louisbourg's Last Decade* Lincoln: University of Nebraska Press, 2008. ProQuest.

Johnson, James M. *Militiamen, Rangers, and Redcoats: The Military in Georgia, 1754–1776*. Macon: Mercer University Press, 1992.

Johnson, Richard R. *Adjustment to Empire: The New England Colonies, 1675–1715*. Rutgers: Rutgers University Press, 1981.

Jones, Charles C., Jr. *Collections of the Georgia Historical Society*. Vol. 4, *The Dead Towns of Georgia*. Savannah: Morning News Steam Printing House, 1878.

Jones, Kenneth R. "A 'Full and Particular Account' of the Assault on Charleston in 1706." *South Carolina Historical Magazine* 83, no. 1 (January 1982): 1–11. JSTOR.

Kinkel, Sarah. *Disciplining the Empire: Politics, Governance, and the Rise of the British Navy*. Harvard: Harvard University Press, 2018. JSTOR.

Lanning, John Tate. "The American Colonies in the Preliminaries of the War of Jenkins' Ear." *Georgia Historical Quarterly* 11, no. 2 (June 1927): 129–55. JSTOR.

Leach, Douglas. *Arms for Empire: A Military History of the British Colonies in North America, 1607–1763*. New York: Macmillan, 1973. Google Play eBook.

———. *Roots of Conflict: British Armed Forces and Colonial Americans, 1677–1763*. Chapel Hill: University of North Carolina Press, 1986. Kindle eBook edition.

Lemisch, Jesse. *Jack Tar vs. John Bull: The Role of New York's Seamen in Precipitating the Revolution*. New York: Routledge, 1997. Kindle eBook edition.

Lenman, Bruce P. *Britain's Colonial Wars, 1688–1783*. New York: Routledge, 2001. Google Play eBook.

Lennox, Jeffers. *Homelands and Empires: Indigenous Spaces, Imperial Fictions, and Competition for Territory in Northeastern North America, 1690–1763*. Toronto: University of Toronto Press, 2017. ProQuest.

Lewis, James. *Neptune's Militia: The Frigate South Carolina during the American Revolution*. Kent: Kent University Press, 1999. EBSCOhost.

Lincoln, Waldo. *The Province Snow "Prince of Orange."* Worcester: Press of Charles Hamilton, 1901.

Lipman, Andrew. *The Saltwater Frontier: Indians and the Contest for the American Coast*. New Haven: Yale University Press, 2015.

Little, Benerson. *The Buccaneer's Realm: Pirate Life on the Spanish Main, 1674–1688*. Washington, DC: Potomac Books, 2007.

———. *Pirate Hunting: The Fight against Pirates, Privateers, and Sea Raiders from Antiquity to the Present*. Washington, DC: Potomac Books, 2010.

Lustig, Mary Lou. *The Imperial Executive in America: Sir Edmund Andros, 1637–1714*. Madison: Fairleigh Dickinson University Press, 2002.

Lyons, Adam. *The 1711 Expedition to Quebec: Politics and the Limitations of British Global Strategy*. New York: Bloomsbury Academic, 2013. Google Play eBook.

Macleod, Malcolm. *French and British Strategy in the Lake Ontario Theatre of Operations, 1754–1760*. Unpublished graduate thesis, University of Ottawa, 1973.

Magra, Christopher. *The Fisherman's Cause: Atlantic Commerce and Maritime Dimensions of the American Revolution*. Cambridge: Cambridge University Press, 2009. Kindle eBook edition.

———. *Poseidon's Curse: British Naval Impressment and Atlantic Origins of the American Revolution*. Cambridge: Cambridge University Press, 2016.

Maier, Pauline. *From Resistance to Revolution: Colonial Radicals and the Development of American Opposition to Britain, 1765–1776*. New York: Alfred A. Knopf, 1973.

Malcomson, Robert. "Not Very Much Celebrated: The Evolution and Nature of the Provincial Marine, 1755–1813." *Northern Mariner* 11, no. 1 (January 2001): 25–37.

———. *Warships of the Great Lakes, 1754–1834*. Annapolis: Naval Institute Press, 2001.

Mancke, Elizabeth. "Negotiating an Empire: Britain and Its Overseas Peripheries, c. 1550–1780." In *Negotiated Empires: Centers and Peripheries in the Americas, 1500–1820*, edited by Christine Daniels and Michael V. Kennedy London: Taylor & Francis Group, 2002. ProQuest.

Marley, David. *Wars of the Americas: A Chronology of Armed Conflict in the New World, 1492 to the Present*. Santa Barbara: ABC-CLIO, 1998.

May, W. E. "Capt. Charles Hardy on the Carolina Station, 1742–1744." *South Carolina Historical Magazine* 70, no. 1 (1969): 1–19. JSTOR.

———. "Captain Frankland's *Rose*." *American Neptune* 26 (1966): 37–62.

McDermott, James. *England & the Spanish Armada: The Necessary Quarrel*. New Haven: Yale University Press, 2005.

McDonald, Kevin P. "Sailors from the Woods: Logwood Cutting and the Spectrum of Piracy." In *The Golden Age of Piracy: The Rise, Fall, and Enduring Popularity of Pirates*, edited by David Head. Athens: University of Georgia Press, 2018.

McLaughlan, Ian. *The Sloop of War, 1650–1763*. Barnsley: Seaforth Publishing, 2014.

Meany, Joseph F. "Batteau and 'Battoe Men': An American Colonial Response to the Problems of Logistics in Mountain Warfare." Manuscript published digitally by New York State Military Museum.

Middleton, Arthur Pierce. *Tobacco Coast: A Maritime History of Chesapeake Bay in the Colonial Era*. Newport News: Mariners' Museum, 1953.

Miles, William R. *The Royal Navy and Northeastern North America, 1689–1713*. Unpublished master's thesis, Saint Mary's University, Halifax, Nova Scotia, 2000. Collections Canada website.

Moses, Norton H. "The British Navy and the Caribbean, 1689–1697." *Mariner's Mirror* 52, no. 1 (1966): 13–40.

Nagel, Kurt. *Empire and Interest: British Colonial Defense Policy, 1689–1748.* Unpublished PhD diss., Johns Hopkins University, 1992.
Nash, Gary. *The Urban Crucible: The Northern Seaports and the Origins of the American Revolution.* Cambridge: Harvard University Press, 1986.
Newbold, Robert Clifford, *The Albany Congress and Plan of Union of 1754.* New York: Vantage Press, 1955.
Norton, Mary Beth. *In the Devil's Snare: The Salem Witchcraft Crisis of 1692.* New York: Vintage Books, 2002. Kindle eBook edition.
Oatis, Steven J. *A Colonial Complex: South Carolina's Frontiers in the Era of the Yamasee War, 1680–1730.* Lincoln: University of Nebraska Press, 2004. EBSCOhost.
O'Connor, Raymond G. *Origins of the American Navy: Sea Power in the Colonies and the New Nation.* Lanham: University Press of America, 1994.
Pares, Richard. *War and Trade in the West Indies, 1739–1763.* London: Frank Cass, 1963.
Paullin, Charles O. *Colonial Army and Navy.* Unpublished manuscript, Charles Oscar Paullin papers, 1931. MSS53033, Library of Congress.
———. *The Navy of the American Revolution: Its Administration, Its Policy and Its Achievements.* Chicago: University of Chicago, 1906.
Pestana, Carla Gardina. *The English Atlantic in an Age of Revolution, 1640–1661.* Cambridge: Harvard University Press, 2007. ProQuest.
Powers, David M. "'Use Dilatory Means': William Pynchon and the Native Americans." Conference Paper Presented at 17th Century Warfare, Diplomacy & Society in the American Northeast. Pequot War website.
Quinn, D. B. "CUMINGS, ARCHIBALD." In *Dictionary of Canadian Biography.* Vol. 2 University of Toronto/Université Laval, 2003.
Rawlyk, George. *Yankees at Louisbourg.* Orono: University of Maine Press, 1967.
Ray, Benjamin C. *Satan and Salem: The Witch-Hunt Crisis of 1692.* Charlottesville: University of Virginia Press, 2015.
Rediker, Marcus, and Peter Linebaugh. *The Many-Headed Hydra: The Hidden History of the Revolutionary Atlantic.* Boston: Beacon Press, 2000.
Reese, Trevor R. "Colonial Georgia in British Policy, 1732–1756." Unpublished diss., University of London, 1955. ProQuest.
———. *Colonial Georgia: A Study in British Imperial Policy in the Eighteenth Century.* Athens: University of Georgia Press, 1963.
———. "Georgia in Anglo-Spanish Diplomacy, 1736–1739." *William and Mary Quarterly* 15, no. 2 (April 1958): 168–90. JSTOR.
Richmond, H. W. *The Navy in the War of 1739–48.* Vol. 3. Cambridge: Cambridge University Press, 1920.
Ritchie, Robert. *Captain Kidd and the War against the Pirates.* Boston: Harvard University, 1986.
Robinson, John, and George Francis Dow. *The Sailing Ships of New England, 1607–1907.* New York: Skyhorse Publishing, 2007.
Robson, Martin *A History of the Royal Navy: The Seven Years War.* London: I. B. Tauris, 2016. EBSCOhost.
Rodger, N. A. M. *The Command of the Ocean: A Naval History of Britain, 1649–1815.* New York: W. W. Norton, 2004.

———. "The Law and Language of Private Naval Warfare." *Mariner's Mirror* 100, no. 1 (2014): 5–16.

———. "The New Atlantic: Naval Warfare in the Sixteenth Century." In *War at Sea in the Middle Ages and the Renaissance*, edited by John B. Hattendorf and Richard W. Unger. London: Boydell & Brewer, 2003. JSTOR.

———. *The Safeguard of the Sea: A Naval History of Britain, 640–1649*. New York: W. W. Norton, 1997.

Rogers, Alan. *Empire and Liberty: American Resistance to British Authority, 1755–1763*. Berkeley: University of California Press, 1974.

———. *Murder and the Death Penalty in Massachusetts*. Amherst: University of Massachusetts Press, 2008.

Rose, Susan. *England's Medieval Navy, 1066–1509: Ships, Men & Warfare*. Montreal: McGill-Queen's University Press, 2013.

Sanders, G. Earl. "Counter-Contraband in Spanish America: Handicaps of the Governors in the Indies." *Americas* 34, no. 1 (1977): 59–72. JSTOR.

Satsuma, Shinsuke. *Britain and Colonial Maritime War in the Early Eighteenth Century: Silver, Seapower and the Atlantic*. London: Boydell & Brewer, 2013. JSTOR.

Selesky, Harold E. *War and Society in Colonial Connecticut*. New Haven: Yale University press, 1990.

Schaffer, Benjamin. "Pirates, Politics, and Provincial Navies: Colonial South Carolina and Its Naval Forces, 1700–1719," Unpublished manuscript, accepted for publication at *South Carolina Historical Magazine*.

Schuler, Jack. *Calling Out Liberty: The Stono Slave Rebellion and the Universal Struggle for Human Rights*. Jackson: University Press of Mississippi, 2009. ProQuest.

Sherman, Richard P. *Robert Johnson: Proprietary & Royal Governor*. Columbia: University of South Carolina Press, 1966.

Sirmans, M. Eugene. *Colonial South Carolina: A Political History 1663–1763*. Chapel Hill: University of North Carolina Press, 1966.

Smith, D. E. Huger. "Commodore Alexander Gillon and the Frigate South Carolina." *South Carolina Historical and Genealogical Magazine* 9, no. 4 (1908): 189–219.

Smith, Gordon Burns. *Morningstars of Liberty: The Revolutionary War in Georgia, 1775–1783*. Milledgeville, GA: Boyd Publishing, 2006.

Smith, Mark M. "African Dimensions." In *Documenting and Interpreting a Southern Slave Revolt*, edited by John K. Thornton. Columbia: University of South Carolina Press, 2005. JSTOR.

Smith, Philip Chadwick Foster. "King George, the Massachusetts Province Ship, 1757–1763: A Survey." In *Seafaring in Colonial Massachusetts, Publications of the Colonial Society of Massachusetts*, vol. 52, edited by Frederick S. Allis Jr., 175–85. Boston: Colonial Society of Massachusetts, 1980. Colonial Society of Massachusetts website.

Snow, Caleb Hopkins. *A History of Boston: The Metropolis of Massachusetts, from Its Origin to the Present Period*. Boston: A Bowen, 1828.

Sosin, Jack M. *English America and the Restoration Monarchy of Charles II: Transatlantic Politics, Commerce, and Kinship*. Lincoln: University of Nebraska Press, 1980.

———. "Review of 1676 [. . .]." *Virginia Magazine of History and Biography* 93, no. 2 (1985): 213–14. JSTOR.
Stanwood, Owen. *The Empire Reformed: English America in the Age of the Glorious Revolution*. Philadelphia: University of Pennsylvania Press, 2013. ProQuest.
———. "Review of Marlborough's America." *William and Mary Quarterly* 71 (July 2014): 484–87. JSTOR.
Starkey, David J. *British Privateering Enterprise in the Eighteenth Century*. Exeter: University of Exeter Press, 1990.
———. "Voluntaries and Sea Robbers: A Review of the Academic Literature on Privateering, Corsairing, Buccaneering and Piracy." *Mariner's Mirror* 97, no. 1 (2011): 127–47.
Steele, Ian K. *The English Atlantic, 1675–1740: An Exploration of Communication and Community*. New York: Oxford University Press, 1986. EPUB.
Stout, Neil R. *The Royal Navy in America, 1760–1775*. Annapolis: Naval Institute Press, 1973.
Swanson, Carl E. *Predators and Prizes: American Privateering and Imperial Warfare, 1739–1748*. Columbia: University of South Carolina Press, 1990.
———. "'The Unspeakable Calamity This Poor Province Suffers from Pyrats': South Carolina and the Golden Age of Piracy." *Northern Mariner/Le Marin Du Nord* 21, no. 2 (2011): 117–42.
Sweeney, Alexander Y. "Cultural Continuity and Change: Archaeological Research at Yamasee Primary Towns in South Carolina." In *The Yamasee Indians: From Florida to South Carolina*, edited by Denise Bossy. Lincoln: University of Nebraska Press, 2018. JSTOR.
Tager, Jack. *Boston Riots: Three Centuries of Social Violence*. Boston: Northeastern University, 2001.
Tapley, Harriet Silvester. *The Province Galley of Massachusetts Bay, 1694–1716: A Chapter of Early American Naval History*. Salem: Essex Institute, 1922.
Taylor, Alan. *American Colonies: The Settling of North America*. New York: Penguin, 2001. Kindle eBook edition.
Truxes, Thomas. "The Breakdown of Borders: Commerce Raiding during the Seven Years' War, 1756–1763." In *Commerce Raiding: Historical Case Studies, 1755–2009*, edited by Bruce A. Elleman and S. C. M. Paine. New Port: Naval War College, 2013.
Vickers, Daniel. "The Northern Colonies: Economy and Society, 1600–1775." In *The Cambridge Economic History of the United States*, vol. 1, edited by Stanley L. Engerman and Robert E. Gallman. Cambridge: Cambridge University Press, 1996.
Watson, Michael. "Judge Lewis Morris, the New York Vice-Admiralty Court, and Colonial Privateering, 1739–1762." *New York History* 78, no. 2 (April 1997): 117–46.
Webb, Stephen Saunders. *The Governors-General: The English Army and the Definition of Empire, 1569–1681*. Chapel Hill: University of North Carolina Press, 1979.
Wilkinson, Clive. *The British Navy and the State in the Eighteenth Century*. London: Boydell & Brewer, 2004. JSTOR.
Williams, Glyndwr. *The British Atlantic Empire before the American Revolution*. London: Taylor & Francis Group, 1980.

Willis, Sam. *The Struggle for Sea Power: A Naval History of American Independence*. New York: W. W. Norton, 2016.
Wilson, David. "Protecting Trade by Suppressing Pirates: British Colonial and Metropolitan Responses to Atlantic Piracy, 1716–1726." In *The Golden Age of Piracy: The Rise, Fall, and Enduring Popularity of Pirates*, edited by David Head. Athens: University of Georgia Press, 2018.
Winfield, Rif, and Stephen S. Roberts. *French Warships in the Age of Sail, 1626–1786: Design, Construction, Careers and Fates*. South Yorkshire: Seaforth Publishing, 2017.
Woodward, Colin. *The Republic of Pirates: Being the True and Surprising Story of the Caribbean Pirates and the Man Who Brought Them Down*. Orlando: Harcourt, 2007.
Wright, Robert K., Jr. *Continental Army*. Washington, DC: Center of Military History, 1983. US Army, Center of Military History website.
Zelner, Kyle F. *A Rabble in Arms: Massachusetts Towns and Militiamen during King Philip's War*. New York: New York University Press, 2009. ProQuest.

Websites/Digital Resources

"Archangel." Three Decks-Warships in the Age of Sail Database.
"Stephen Hopkins, 1707–1785." US House of Representatives.

Index

Adams, John, 138–39
Adams, Samuel, 106, 109, 139
Admiralty of England (and later Great Britain), 10–12, 14–15, 19–20, 46, 53–54, 60, 73, 83–84, 93, 97, 100–102, 105–6, 111–12, 121–22, 124, 132
Albany Congress (1755), 114–15
Alden, John, Massachusetts provincial navy captain, 30, 42–43
Alexandria Congress (1755), 115
American Revolution. *See* Revolutionary War (1775–1783)
Andros, Sir Edmund, governor of Dominion of New England: coup against, 29–31; provincial navy of, 30–31, 40–42
Anglo-French War (1627–1629), 17
Anglo-Spanish War (1585–1603), 15–16
Anglo-Spanish War (1625–1630), 17
Anglo-Spanish War (1654–1660), 19
Anglo-Spanish War (1727–1729), 49, 65–66, 73
Anson, Baron George, Royal Navy Admiral, 66, 83, 101, 105, 110, 112
Antigua. *See* Leeward Islands
Auchmuty, Robert, Massachusetts vice admiralty judge, 103–4

Bahamas: governors of, 70; as a pirate nest, 70; privateers from, 70; provincial naval forces of, 70; and the War of the Quadruple Alliance, 70
Barbados: governors of, 68; legislature of, 25, 44–45; privateers from, 33, 44; provincial naval forces of, 25, 33, 44–45, 68; Royal Navy station ships, 19, 107
Barbary Corsairs, 17
Barnwell, John, South Carolina provincial navy captain, 59–60
Berkeley, Sir William, governor of Virginia, 24–25
Bermuda, provincial naval forces of, 120
Bernard, Sir Francis, governor of Massachusetts, 117, 129–30
Blackbeard/Edward Teach, pirate captain, 62, 69–70
Board of Trade, 42, 46, 53, 55, 61, 65, 68, 70–72, 85, 91, 99, 110–11, 113–14, 119
Bonnet, Stede, pirate captain, 61–63
Braddock, Edward, British general, 115
Bradstreet, John, Massachusetts provincial military officer, 116
Bradstreet, Simon, governor of Massachusetts, 31, 40–41
British Army, 65, 90, 94–96, 106, 116, 119, 123–24, 126
British West Florida, Loyalist provincial navy of, 14
buccaneers, 25–26, 29
Bull, Dixie, pirate, 8–9
Bull, William, lieutenant governor of South Carolina, 89–92, 94–95
Canada: Anglo-American joint capture of Fortress Louisbourg during King George's War (1745), 100–106;

Anglo-American joint capture of Newfoundland (1762), 117; Anglo-American joint sieges of Port Royal and Quebec during Queen Anne's War (1707, 1710, 1711), 37–38; borderlands maritime disputes between Anglo-American and Franco-Indigenous forces in the 1710s–20s, 51–56; British capture of Fortress Louisbourg (1758), 122–23; and Father Le Loutre's War (1749–1755), 111–12, 121; Fortress Louisbourg, 56, 84, 86, 98, 100–106, 108, 110–12, 122–23, 125–27, 132, 138; Halifax, Nova Scotia, 110–12, 125; New England assault on Quebec (1690), 35–37; New England capture of Port Royal, Nova Scotia (1690), 35; Newfoundland fisheries, 67; provincial naval forces of Nova Scotia, 53, 109–10; Royal Navy station ships in, 110–12

Codrington, Christopher, governor of Leeward Islands, 32–33

Colville, Alexander, Royal Navy rear admiral and Lord Colville, 117, 133

Commonwealth Navy, 19–20

Connecticut: King William's War and, 35; legislature of, 22; as part of the Dominion of New England, 29; Pequot War and, 21–24; provincial naval forces of, 22–23, 98, 117; and Queen Anne's War, 37–38; and the Seven Years' War, 117; and the War of Jenkins' Ear/King George's War, 101

Continental Congress, 138–40

Continental Navy, 114, 138–40

Cornewall, James, Royal Navy captain, 57–58

court cases/trials, courts of admiralty, 12, 103–6

Cromwell, Oliver, Lord Protector, 17–19, 25

Cuba: Anglo-American attack on during the War of Jenkins' Ear, 84; joint Anglo-American capture of Havana (1762), 123–24; as a Spanish naval base, 71–72, 89

Dickinson, John, 139

Drake, Sir Francis, 15–16

Dudley, Joseph, governor of New Hampshire/Massachusetts, 29, 33–34, 44

Dummer, William, lieutenant governor of Massachusetts, 57

Dummer's/Father Rale's War (1722–1725). *See* Canada; Wabanaki Confederacy

Durell, Thomas, Royal Navy captain, 53, 55, 57

Dutch Empire, 19, 21–22, 24–25

Ellis, Sir Henry, governor of Georgia, 119

English Civil Wars, 18–19. *See also* Commonwealth Navy; Cromwell, Oliver

Father Le Loutre's War (1749–1755), 111–12, 121

Florida: joint Anglo-American siege of St. Augustine (1740), 87–90; South Carolina attack on St. Augustine (1702), 36–38

France/French Empire, military forces of, 38–39, 41–43, 51–56, 98, 112, 121–24

Franklin, Benjamin, 2, 54, 87, 114, 139

Gadsden, Christopher, 139

George, John, Royal Navy captain, 29, 40–41

Georgia: British military colonization of, 65, 82; provincial naval forces of, 91, 94–96, 109, 118–19, 142–43; Royal Navy station ships, 91; state navy during the Revolutionary War, 140–41; and the War of Jenkins' Ear (1739–1748), 82–96

Gerry, Elbridge, 140

Glorious Revolution, 30–31, 41–42

governors. *See individual names*

Great Lakes, provincial navies on, 115–16

guarda costas (Spanish), 48, 67–74, 82

Hallowell, Benjamin, provincial navy captain, 117, 134, 138, 141–42

Hamar, Joseph, Royal Navy captain, 92, 94–95

Hamilton, Lord Archibald, governor of Jamaica, 69–70

Hancock, John, founding father, 134

Hardy, Charles, Royal Navy captain, 93–96

Hawkins, Sir John, 15

Hildesley, John, Royal Navy captain, 64–65

Hopkins, Stephen, Rhode Island statesman, 132, 139

imperial crisis of the 1760s–70s, 126–38

impressment, of ships or mariners: by provincial authorities/navies, 2, 5–6, 21,

INDEX 209

24–25, 44–45, 54–55, 92, 126; resistance to, 10–11, 57, 88–89, 107–9, 125–27, 136–37; by royal authorities/the Royal Navy, 4, 10–11, 13, 15–17, 45–46, 57, 88–89, 92, 107–9, 125–27, 136–37
Iroquois Confederacy (Indigenous), 37–38, 113

Jacobites, 69, 72
Jamaica: and the Anglo-Spanish War (1727–1729), 73–74; English capture and colonization of, 19; governors of, 25–26, 71–73; legislature of, 71–73; provincial naval forces of, 71–73, 120; privateers from, 69–71; Royal Navy facilities in, 67, 107; Royal Navy station ships and squadrons in, 19, 26, 28–29, 69–74, 107, 120; and the War of the Quadruple Alliance (1718–1720), 70–71
Johnson, Sir Nathaniel, governor of South Carolina, 38–39
Johnson, Robert, governor of South Carolina, 62–63, 65
Joyner, John, provincial and state navy captain of South Carolina, 140–41

Keppel, Augustus, Royal Navy admiral, 115, 126
King George's War (1744–1748), 96–110. *See also* War of Jenkins' Ear
King Philip's War (1675–1676), 29, 43
King William's War (1689–1698), 27–28, 31–36
Knowles, Charles, Royal Navy commodore, 108–9

Lawes, Sir Nathaniel, governor of Jamaica, 71–72
Leeward Islands: early provincial naval forces, 32–33; provincial naval forces of Antigua, 107; Royal Navy facilities in Antigua, 67, 107; Royal Navy station ships at Antigua, 107; Royal Navy station ships at St. Kitts, 19
legislatures. *See individual colonies*
Little, Thomas, Massachusetts provincial navy captain, 57–58
London, England, 10

Maine: and Dummer's/Father Rale's War (1722–1725), 55–56; King William's War (1689–1698), 34, 42–43; as part of the Dominion of New England, 29–30; and Queen Anne's War (1702–1713), 33–34
marines: British, 99, 119; provincial, 23, 56, 94–96, 109
Maryland: provincial marines from, 94
Massachusetts: early colonization of, 21; legislature of, 31, 34, 54–58, 98–100, 102, 129–30; militia and provincial troops of, 21, 44, 55–56, 106; as part of the Dominion of New England, 29; privateers from, 140; provincial naval forces of, 8, 23–26, 40–44, 46–47, 53–56, 97–106, 108–9, 116–17, 123, 126, 129–31; Royal Navy station ships in, 40–41, 46–47, 51–53, 57–58, 98–99, 107–9; state navy during the Revolutionary War, 138–40
Mead, Samuel, Royal Navy captain, 60–61
merchant ships and trade, 15–16, 24, 62–63, 84–86, 89, 93, 118, 133
monarchs of England (and Great Britain after 1707): Anne, 37, 44; Charles I, 17–19, 21; Charles II, 19–20, 24–25; Edgar II, 9–10; Edward the Confessor, 10; Elizabeth I, 15–16; George I, 62, 65; George II, 83, 96–97, 100, 108; Henry V, 11; Henry VI, 11; Henry VIII, 14; James I, 16–17; James II, 27, 30–31; John, 12; William III, 27, 30–31, 41–42, 46, 140
monarchs of France: Louis XIV, 37, 82; Louis XV, 112; Philip II, 12
monarchs of Spain: Charles III, 123; Filip VI, 123
Montagu, John, Earl of Sandwich, 83, 105, 112, 122
Montagu-Dunk, George, Earl of Halifax, 110–11, 113–14, 117, 119
Moore, James, governor of South Carolina, 36–37
Morgan, Sir Henry. *See* buccaneers

Narragansett people, Indigenous, 22–23
Newfoundland. *See* Canada
New Hampshire: as part of the Dominion of New England, 29; provincial naval forces of, 101; Queen Anne's War, 38; and the War of Jenkins' Ear/King George's War, 101

New York: English capture of New Amsterdam, 19; and King William's War, 35; as part of the Dominion of New England, 29, 41–42; provincial naval forces of, 86, 100; and Queen Anne's War, 37; and the Seven Years' War, 113–14, 116, 126; and the War of Jenkins' Ear/King George's War, 86, 101

North Carolina, capture of the pirate Stede Bonnet (1718), 62

Nova Scotia. *See* Canada

Oglethorpe, James Edward, Georgia military general and statesman, 82–83, 85–92, 94–96, 109, 130–31

Otis, James, Massachusetts statesman, 129–30

Parliament of England (Great Britain after 1707): acts/laws passed by, 45, 93, 97, 99, 107; faction during the English Civil War, 17–20; funding of provincial navies, 87, 96, 109–10, 125; institutional power of, 11, 17, 93

Pearce, Vincent, Royal Navy captain, 87–90

Pease, Samuel, provincial navy captain, 40–41

Pelham, Henry, British prime minister, 112

Pelham-Holles, Thomas, Duke of Newcastle, 90–91, 101–2, 115, 121–22

Pennsylvania: and Queen Anne's War (1702–1713), 37; legislature of, 118; provincial naval forces of, 1–2, 86, 118; Quaker hostility to provincial navies, 114; and the Seven Years' War (1756–1763), 118; and the War of Jenkins' Ear (1739–1748), 86

Pepperell, Sir William, military officer of Massachusetts, 102, 104

Pepys, Samuel, 19

Pequot people, Indigenous, 21–23

Pequot War (1636–1638), 21–22

Philipps, Richard, governor of Nova Scotia, 53

Phips, Sir William, governor of Massachusetts, 35–37, 46

piracy: Golden Age of Piracy (early 18th-century), 49, 61–63, 69–72; individual cases of piracy, 8, 34, 61–63, 40–41, 54–55, 69–72; late seventeenth century piracy in the West Indies, 29–30; legal definitions and cases relating to, 11–12, 62–63, 69–70; as a political label for one's enemies, 134

Pitt, William, British statesman, 112, 117, 122–24, 127

Pound, Thomas, pirate/former provincial navy captain, 40–41

Pownall, Thomas, governor of Massachusetts, 114, 117

privateering/privateers: American Revolutionary War privateers, 140; Anglo-American/British, 15–19, 25–26, 30, 32–33, 39, 69–71, 93, 97–98, 102, 118–21, 126; French, 1–2, 90, 101, 107–8, 117–18, 122, 135; historical origins of, 4, 12–15; historiographical treatment of, 3–7; laws relating to and legality of, 69–70, 97–98, 99–100, 103–6; similarities with provincial navies, 71, 93, 99–100, 103–6; Spanish, 1–2, 65–68, 86, 90, 93, 107. *See also guarda costas*

Providence Island Company, 18–19. *See also* privateering/privateers; Puritans

provincial navies: British governmental policies toward, 53–54, 98–106, 109–10; definition of, 6–7; emergency fleets, 24–25, 38–39, 54–55, 62–63, 73–74, 118–19, 133–35; financing/costs of, 23–25, 32–34, 36, 44–45, 55, 61–63, 68, 71–72, 87–88, 96, 116–18, 120; involvement in diplomacy and conflicts with Indigenous peoples, 34, 42–43; joint operations between colonies, 23–24, 35–36, 86–92, 98, 101; joint operations with Royal Navy, 30, 37–38, 45, 54–55, 73–74, 86–92, 94–96, 100–107, 110, 113, 115–16, 118–19; maintenance of vessels, 44–45, 142; manning of, 25, 44, 87–88, 92–93, 98–99 (*see* impressment); pirate hunting operations of, 34, 54–58, 61–63, 70; semi-permanent or standing forces, 23, 25, 36, 44, 64, 68–69, 71–72, 74–75, 92–93, 98–99, 109–10, 117–18; social ramifications of provincial naval costs, 40–45, 72, 125, 129–30; supplying of, 25, 44, 71; tensions with Royal Navy, 38, 45–47, 57–58, 103–6, 108. *See also* scout boats

Puritans, 18–19, 22, 43

Quebec. *See* Canada

Queen Anne's War (1702–1713), 27–28, 36–39, 44–47

INDEX 211

Raleigh, Sir Walter, 15–17
Randolph, Edward, 29
Revolutionary War (1775–1783). *See also* Continental Navy; *and individual state navies*
Rhett, William, colonel and vice admiral of South Carolina provincial naval forces, 39, 62–63, 65
Rhode Island: as part of the Dominion of New England, 29; provincial naval forces of, 54, 98, 101, 118, 132; Queen Anne's War, 38; resistance to British trade restrictions during the imperial crisis, 133–35; and the War of Jenkin's Ear/King George's War, 98, 101, 132
Rich, Robert, Earl of Warwick, 18–19
Rogers, Woodes, privateer and governor of the Bahamas, 70
Rous, John, provincial navy, privateer, and Royal Navy captain, 93, 102, 108, 110–12, 121
Royal Navy: expansion of in 1730s, 83–84; facilities in North America and the West Indies, 29, 67, 83–84; medieval roots of, 9–11; North American Station, 101–2; pirate hunting by, 29; provincial critiques of, 69, 71–73, 88–90, 92–96, 103–9, 119; role during the imperial crisis of the 1760s–70s, 131–37; social status of officers, 18; types of ships deployed as station ships in colonies, 28. *See also* Commonwealth Navy; *and individual vessels*
Russell, John, Duke of Bedford, 83, 101–2, 112, 122

sailors: payment of (including insurance/prize money allotted to), 24–25, 33–34, 36, 44, 57–58, 62–63, 96, 98–99, 102–6, 125; political and social actions of, 10–11, 36, 135–37; provisioning of, 10, 23, 25, 32, 41, 44, 54–56, 88; social backgrounds and diversity of, 71, 74–75, 87–88, 92; voluntary enlistment by, 39, 57, 88, 98–99, 119, 135. *See also* impressment
Salem witchcraft trials, 42–43
scout boats (South Carolina and Georgia), 59–60, 63–64, 67, 82–83, 86–87, 94, 96, 109, 116, 119, 124, 130–31, 140, 142, 146

Seven Years' War (1756–1763), 111–32
Shirley, William, governor of Massachusetts, 98, 99, 101–2, 108–10, 115–17
Shute, Samuel, governor of Massachusetts, 54–56
slavery: African, 15, 44, 48, 59–60, 71–72, 88, 91–92, 107; Indigenous, 59; resistance to, 60–61, 63, 73, 91
smuggling, 28, 48, 53, 68, 126, 133–35
Southack, Cyprian, Massachusetts provincial navy captain, 33–34, 37, 44
South Carolina: and the Anglo-Spanish War (1727–1729), 65–67; governors of, 36, 62–64; legislature of, 36, 87–91, 109; militia and provincial troops of, 60; provincial naval forces of, 36, 59–67, 88–96, 106, 118–19, 142–43; provincial naval warfare and political instability, 61–65; and Queen Anne's War, 38–39; Royal Navy station ships, 64–65, 90–96, 118; state navy during the Revolutionary War, 140–41; and the War of Jenkins' Ear, 82–96; and the War of the Quadruple Alliance (1718–1720), 64
Spain/Spanish Empire: naval forces of, 39, 48, 89, 90–91, 95–96, 121; overseas exploration by, 14–15. *See also guarda costas*
Spanish Armada, 16
St. Kitts. *See* Leeward Islands

Third Anglo-Dutch War (1672–1674), 23–25
Trott, Nicholas, South Carolina justice, 62–63
Tyng, Edward, provincial navy captain, 84, 109

Vane, Charles, pirate captain, 62, 70
Vernon, Edward, Royal Navy admiral, 72
vessels: *Abercrombie* (privateer brigantine), 118; *Beaufort Galley* (provincial navy galley), 92; *Boston Packet* (provincial navy brigantine), 104–6; *Carolina* (provincial navy scout boat), 82; *Charles Town* (provincial navy galley), 84, 92, 94–95; *Constant Jane* (provincial navy sloop), 45; *Defence* (various provincial navy sloops), 98, 117, 132; *Delicia* (privately-owned armed ship), 70; *Faulcon* (provincial navy sloop), 94; *Gaspee* (British revenue schooner),

134–35; *Henry* (provincial navy sloop), 62–63; HMS *Archangel* (Royal Navy hired sloop), 42; HMS *Arundel* (Royal Navy frigate), 119; HMS *Chester* (Royal Navy frigate), 104–5; HMS *Conception Prize* (Royal Navy frigate), 46; HMS *Dragon* (Royal Navy frigate), 38; HMS *Elizabeth* (Royal Navy frigate), 24; HMS *Flamborough* (Royal Navy frigate), 64–65, 88, 92, 94–95; HMS *Hawk* (Royal Navy sloop), 86, 93; HMS *Hunter* (Royal Navy sloop), 125; HMS *Kingfisher* (Royal Navy frigate), 29; HMS *Ludlow Castle* (Royal Navy frigate), 57; HMS *Nonsuch* (Royal Navy frigate), 46; HMS *Oxford* (Royal Navy frigate), 26; HMS *Phoenix* (Royal Navy frigate), 90; HMS *Prince Royal* (unrated Royal Navy warship), 17; HMS *Rose* (various Royal Navy frigates), 29, 31–32, 41, 51, 94; HMS *Rye* (Royal Navy frigate), 93, 95; HMS *Seahorse* (Royal Navy frigate), 54–55; HMS *Sheerness* (Royal Navy frigate), 57–58; HMS *Shirley*, 102, 108; HMS *Sovereign of the Seas* (Royal Navy ship of the line), 18; HMS *Squirrel* (various Royal Navy frigates), 83, 133; HMS *Success* (Royal Navy frigate), 60–61; HMS *Sunderland* (Royal Navy frigate), 104–5; HMS *Tartar* (Royal Navy frigate), 88–90; HMS *Wager* (Royal Navy frigate), 108; *John & Abiel* (provincial navy brigantine), 44; *King George* (provincial navy frigate), 117, 123, 125, 134, 141–42; *Larke* (provincial navy brigantine), 45; *Loyal Heart* (provincial navy sloop), 57–58; *Madeira* (provincial navy brigantine), 45; *Mary* (provincial navy sloop), 30–31, 35, 40–43, 46; *Massachusetts* (provincial navy frigate), 84, 101, 109; *Massachusetts* (provincial navy sloop), 129–30; *Massachusetts* (state navy brigantine), 142; *Notre Dame de Deliverance* (prize ship), 104–6; *Palmer* (provincial navy sloop), 65–66; *Pearl* (provincial navy schooner), 90; *Pennsylvania Frigate* (provincial navy frigate), 118; *Prince George* (provincial/Georgia state navy scout boat), 109, 130–31, 142; *Prince of Orange* (provincial navy snow), 84, 98–99; *Prince of Wales* (provincial navy snow), 117; *Province Galley* (two provincial navy galleys), 31–34, 37–38, 44; *Resolution* (various provincial navy sloops), 30, 40–41, 103–4, 108; *Sampson* (privateer vessel), 126; *Samuel* (provincial navy brigantine), 30; *Sarah* (provincial navy sloop), 30; *Sea Nymph* (provincial navy sloop), 62–63; *Six Friends* (provincial navy ship), 35; *South Carolina* (state navy frigate), 141; *Speedwell* (provincial navy ketch/sloop), 30, 41–42; *St. John* (British revenue schooner), 133–34; *St. Philip* (provincial navy sloop), 94; *Success* (provincial navy ship), 95; *Tartar* (provincial navy brigantine), 117–18; *Tartar* (provincial navy sloop), 84, 98, 132; *Tryal* (provincial navy sloop), 119; *The Two Friends* (merchant ship), 103–4; *Walker* (provincial navy schooner), 94; *William Augustus* (provincial navy schooner), 53

Vetch, Samuel, 37–38

Virginia: Dutch attack on, 23–25; early colonization of, 20; provincial marines from, 94; Royal Navy station ships, 19, 24

Wabanaki Confederacy (including Indigenous Abenaki, Mi'kmaq, and Maliseet peoples): alliance with French, 53; conflicts with Anglo-Americans, 51–53, 110–11

Walpole, Robert, prime minister of Great Britain, 50, 67–68, 74, 82–83, 87

Wanton, Joseph, Sr., governor of Rhode Island, 134–35

War of Jenkins' Ear/King George's War, 82–111

War of the Quadruple Alliance (1718–1720), 49, 64, 71

Warren, Sir Peter, Royal Navy vice admiral, 86, 89, 100–105, 108, 132

Washington, George, 138, 142

Winthrop, John, governor of Massachusetts, 8

women and provincial navies, 55–56

Yamasee people, Indigenous, 49, 59, 61, 65–67

Yamasee War (1715–1720s), 59–61